SpringerWienNewYork

Ferdinand Cap

# Tsunamis and Hurricanes

## A Mathematical Approach

**Emer. Univ.-Prof. Ferdinand Cap**
Innsbruck, Austria


Softcover re-print of the Hardcover 1st edition 2006

SpringerWienNewYork is part of
Springer Science+Business Media
springer.com

Typesetting: Camera ready by the author
(Mathematica fonts by Wolfram Research, Inc.)

Printing: Strauss GmbH, 69509 Mörlenbach, Germany

Printed on acid-free and chlorine-free bleached paper
SPIN: 11607502

With 21 Figures

ISBN-978-3-211-33159-0 (eBook)

ISBN-10 3-7091-1737-1 SpringerWienNewYork
ISBN-13 978-3-7091-1737-8 SpringerWienNewYork

DOI 10.1007/ 978-3-211-33159-0

# Preface

Tsunamis and hurricanes are natural catastrophes which can make considerable material damage and personal harm to humans. Any possibility to describe these phenomena and to find methods of predictability of any kind seem therefore to be of interest not only for meteorologists, but also for governments, evacuation plans or the insurance industry etc. There may now exist a chance to satisfy these needs, if a tsunami wave equation could be found and solved. Seismic waves in the Earth's crust propagate faster (4–6 km/sec) than tsunamis (100–900 km/h). This speed difference allows an early warning time of up to a few hours, depending on the location of the earthquake or underwater explosion. If a tsunami wave equation and its solutions were known, even a guess of the tsunami crest height might be possible and useful.

In this book a mathematical approach to tsunami wave equations is presented. To the author's knowledge some of the tools and computer codes presented here have never been applied on tsunamis and hurricanes. Some of the calculations in this book are based on the PREISWERK-LANDAU equivalence principle between gasdynamics and hydrodynamics as well on the BECHERT-MARX linearization method using the mass variable transformation. Other tools used here are similarity transformations and the program packet Mathematica by WOLFRAM. Although knowledge of these codes is not necessary for the use of this book, it may however help to understand some calculations and the reader may acquire some knowledge of this program.

The LIE-series method to solve differential equations are also mentioned or Codes developed by NOAA and other organisations.

The author thanks his colleague H. PICHLER of the Meteorology Department of Innsbruck University for a critical reading of the manuscript and for many useful suggestions. The author thanks also his son Clemens of the Institute for Informatics, University of Rostock, for some hints and especially his wife Theresia for providing the typed version of this book. With endless patience, interest and engagement she brought the countless different versions of the often poorly handwritten manuscript into professional format using the computer program LaTeX.

Innsbruck, Austria, September 2006

# Contents

# 1. Introduction to wave physics

## 1.1 Types of waves

A wave is a disturbance of a physical quantity which propagates from one location in space to another point. Mathematically a wave is described in terms of its strength, called amplitude, and how the amplitude varies with both space and time. The description of the wave amplitude is given by the *general solution* of the appropriate *wave equation.* These equations are linear or nonlinear partial differential equations, depending on the type of the physical quantity. The perturbed physical quantity may be a property of a medium like a fluid, a gas or an electromagnetic or another field. In quantum mechanics complex waves are described by the SCHROEDINGER wave equation. The appropriate wave equation is defined by the physical phenomenon, be it a displacement in a medium like water or a physical field or quantity.

If the solution of a wave equation satisfies the equation and some boundary and/or initial condition it is called a *partial solution.* Exterior forces like gravity, wind or earthquakes have a decisive influence on the solutions, whereas intrinsic properties of the medium like surface tension (capillarity), elasticity, inertia or viscosity are taken into account by the appropriate wave equation.

The various types of waves in water may be classified according to the mechanism exciting the wave or according to the appearance and behavior of the waves. Furthermore, linear waves may be classified according to their wave length $\lambda$, the *propagation speed* $c$ and their frequency $\nu$. Linear waves satisfy a linear wave equation. If the coefficients in the wave equation are constants, then the wave equation reads

$$c^2 \Delta u(x, y, z, t) = u_{tt}, \tag{1.1.1}$$

where $\Delta$ is the LAPLACE-*operator* $u_{xx} + u_{yy} + u_{zz}$. If one considers a one-dimensional problem, the general solution of (1.1.1) may be a *travelling wave*

$$u(x, t) = f(x + ct) + g(x - ct), \tag{1.1.2}$$

where $f$ and $g$ are arbitrary functions, or a sinusoidal oscillatory wave of the form

$$u(x, t) = A \sin 2\pi \left( \frac{t}{\tau} - \frac{x}{\lambda} \right). \tag{1.1.3}$$

The wave length $\lambda$ is the horizontal distance between successive crests of the wave, the amplitude $A$ is the distance from the wave surface at rest to the crest and $\tau = 1/\nu$ is the wave period. Then the relation holds

$$c = \lambda/\tau = \lambda\nu = \lambda\omega/2\pi = \omega/k. \tag{1.1.4}$$

As we will see later, these equations (1.1.1)–(1.1.4) are valid for sinusoidal (linear) waves only. Using the concept of the *wave number* $k = 2\pi\nu/c = 2\pi/\lambda$, one may write (1.1.3) in the form

$$u(x,t) = A\cos(2\pi\nu t - kx + \varphi) = A\cos(\Phi), \tag{1.1.5}$$

where $\varphi$ is an arbitrary phase constant (phase angle $\Phi$ in radians). A group of waves of various frequencies $\nu_n$ is described by

$$u(x,t) = \sum_n A_n \cos\left[\omega_n\left(t - \frac{x}{c}\right) + \varphi_n\right], \tag{1.1.6}$$

where $\omega_n = 2\pi\nu_n$. Here $\nu_n$ is the frequency of the $n$-th component of the group of waves. The superposition of waves of different wave length with phases such that the resultant amplitude is finite over a small region is called a *wave packet.* The velocity of energy flow in a propagating wave packet is called *group velocity* $c_g$. The propagation speed $c(\nu)$ of a simple harmonic (sinusoidal) wave is also called *phase velocity.* Waves exhibiting *dispersion* show a dependence of the phase velocity $c$ on the frequency or the wave length.

A wave of frequency $\nu_n$ has a phase velocity $c(\nu_n)$ according to (1.1.4) so that also the wave number $k$ depends on the frequency. If the phase $\Phi$ in (1.1.5) is constant then we find the locations with the phase for various times. For constant $= 0$ one has

$$k\mathrm{d}x - \omega\mathrm{d}t = 0, \quad \text{or} \quad \frac{\mathrm{d}x}{\mathrm{d}t} = \frac{\omega}{k} = c. \tag{1.1.7}$$

$c$ is the phase velocity of a monochromatic wave of frequency $\nu$. Considering now a group of waves (1.1.6), we now may define a *group velocity* $c_g$ by

$$c_g = \frac{\mathrm{d}\omega}{\mathrm{d}k}. \tag{1.1.8}$$

If there is no *dispersion*, if thus the phase speed $c$ does not depend on the frequency (on the wave number), one obtains $\omega = ck$ or $\mathrm{d}\omega = c\mathrm{d}k$ and the phase velocity $c$ is equal to the group velocity. The fact is however, that all water waves exhibit dispersion (see chapter 3).

In the case of dispersion we have to expand

$$c_g = \frac{\mathrm{d}\omega}{\mathrm{d}k} = c + k\frac{\mathrm{d}c}{\mathrm{d}k} = c - \lambda\frac{\mathrm{d}c}{\mathrm{d}\lambda}. \tag{1.1.9}$$

For *normal dispersion* defined by $\mathrm{d}c/\mathrm{d}\lambda > 0$ the group velocity $c_g$ is smaller than the phase velocity $c$. For *anomalous disperion* ($\mathrm{d}c/\mathrm{d}\lambda > 0$) one has $c_g > c$. Both cases are realized by water waves.

Wave packets are defined by a group of waves of the type (1.1.6) with continuously varying frequencies neighboring $\nu_0$ or wave numbers $k_0$. Using the well known expression $\exp(ip) = \cos p + i \sin p$, we thus may replace (1.1.6) by the FOURIER *integral*

$$u(x,t) = \int_{k_0-\varepsilon}^{k_0+\varepsilon} A(k) \exp(i[kx - \omega t])\mathrm{d}k, \tag{1.1.10}$$

which describes a *wave packet.* ($\varepsilon$ is a very small constant quantity). Using $kx - \omega t = k_0 x - \omega_0 t + (k - k_0)x - (\omega - \omega_0)t$ one obtains

$$u(x,t) = \exp(i[k_0 x - \omega_0 t]) \int_{k_0-\varepsilon}^{k_0+\varepsilon} A(k) \exp\left(i[k - k_0]x - [\omega - \omega_0]t\right) \mathrm{d}k. \tag{1.1.11}$$

To obtain the propagation speed of the phase, we find from (1.1.7)

$$\frac{\mathrm{d}x}{\mathrm{d}t} = \frac{\omega - \omega_0}{k - k_0} = \frac{\Delta\omega}{\Delta k} \approx \frac{\mathrm{d}\omega}{\mathrm{d}k} = c_g. \tag{1.1.12}$$

A wave packet propagates with the group velocity $c_g$.

Considering now the mechanisms and forces exciting waves in media we may mention:

1. *tidal waves*, excited by the tides and thus by the combined action of the gravity of Earth and Moon,

2. *surface waves* due to the capillarity of water and excited by wind and pressure differences, see section 3.3,

3. *gravity waves* due to gravity alone and appearing as *breakers*, *surge* etc, section 3.2,

4. surface waves due to the combined action of surface tension (capillarity) and gravity excited by wind, producing *ripples*,

5. *wave packets* like *jumps*, *solitary waves*, *solitons*, *seiches*, *edge waves*, *shallows*, *swells*, *tsunamis* etc,

6. waves connected with the viscosity of water.

On the other hand, a classification due to the wave form is of interest. There exist waves whose shape does not change like *sinosoidal* or *cnoidal* and *snoidal* waves (described by elliptic functions) and there exist waves which change their shape slightly or enormously, for instance by compression or expansion of the medium carrying the waves, by nonlinear effects etc. Wave deformation may also depend on the depth of the lake or ocean. It may also be due to the dependence of the wave velocity both on the amplitude of the wave and/or the depth of the liquid. A special type of waves is given by *internal waves.* These are particular waves in the ocean or in a lake occurring at the interface of two layers of water of different temperature and therefore density. Waves in the atmosphere consider special attention due to the compressibility of the medium and of the effects of Coriolis force on the rotating Earth. *Periodic waves* are defined by $u(x, y, z, t) = u(x, y, z, t + \tau)$ and *oscillatory waves* have a change in sign after a half period.

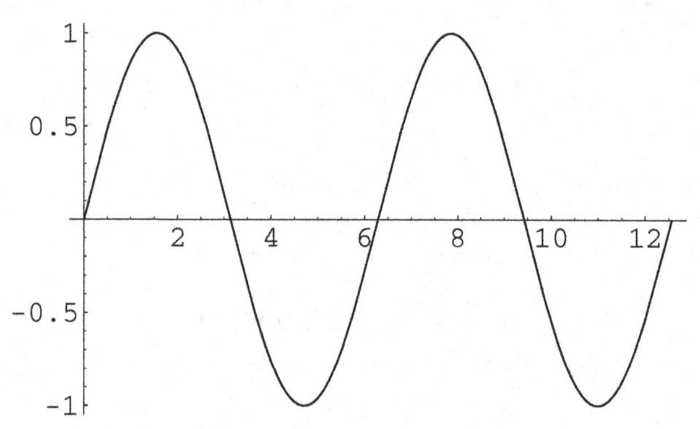

Fig. 1.1. Oscillatory harmonic wave

Figure 1.1 has been produced by the program packet Mathematica [1.1] using the command

```
Plot[Sin[x],{x,0,4*Pi}]
```
(1.1.13)

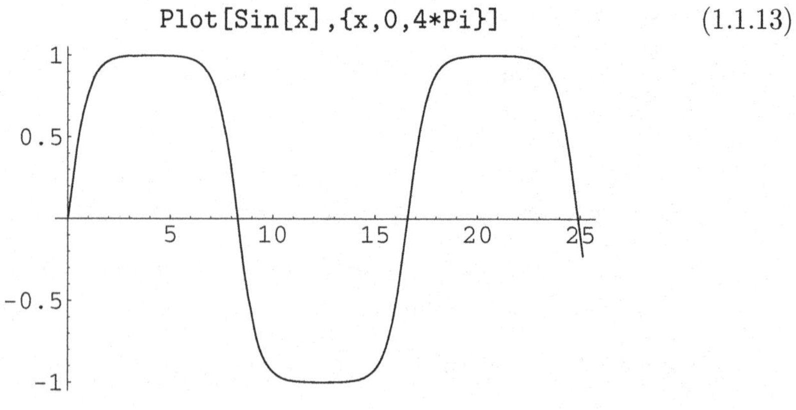

Fig. 1.2. Nonlinear nonharmonic oscillatory periodic wave

Although knowledge and use of the program packet Mathematica are not necessary to fully understand all calculations presented in this book it might help to execute some calculations. Equation (1.1.13) describes a sinusoidal monochromatic wave (1.1.5). Here the replacements $2\pi\nu t - kx + \varphi \to x, A = 1$ have been made. This replacement corresponds to a translation to a co-moving frame (wave frame). Figure 1.2 shows a stable nonlinear nonharmonic oscillatory periodic wave (snoidal or *cnoidal wave*) produced by

```
Plot[JacobiSN[t,0.996],{t,0,8*Pi}]
```
(1.1.14)

Here the replacement has been $2\pi\nu t - kx + \varphi \to t$.

A periodic wave may be but must not be oscillatory. An example is given in Fig. 1.3.

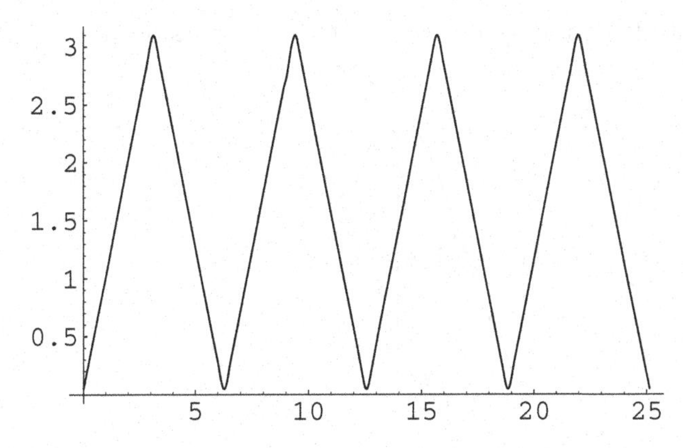


Fig. 1.3. Periodic nonoscillatory wave

This graph has been produced by plotting the FOURIER *series*
```
y[t_]=Pi/2.-4*Sum[Cos[n*t]*n^(-2),{n,1,31,2}/Pi
```

$$y(t) = \frac{\pi}{2} - \frac{4}{\pi}\left(\cos t + \frac{\cos 3t}{3^2} + \frac{\cos 5t}{5^2} + \frac{\cos 7t}{7^2} + ...\right) \tag{1.1.15}$$

and `Plot[y[t],{t,0,8*Pi}]` A periodic oscillatory wave may be represented by

$$y(t) = \frac{4}{\pi}\left(\sin t - \frac{\sin 3t}{3^2} + \frac{\sin 5t}{5^2} - \frac{\sin 7t}{7^2} + ...\right), \tag{1.1.16}$$

see Fig. 1.4.

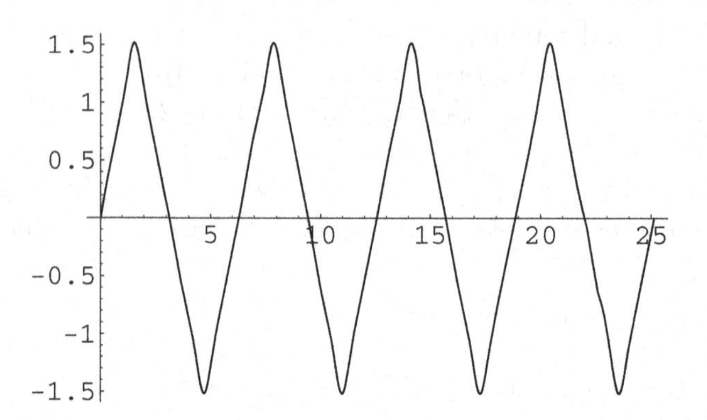


Fig. 1.4. Periodic nonharmonic oscillatory wave

There exist also nonperiodic nonoscillatory waves occurring in water, see Fig. 1.5.

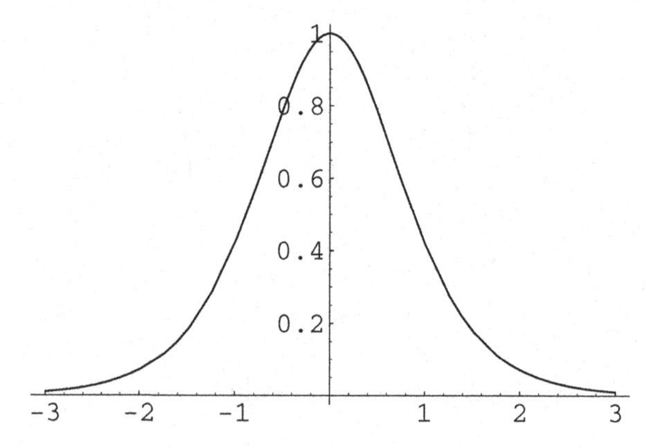


Fig. 1.5. Nonperiodic nonoscillatory wave (a *soliton)*

Figure 1.5 has been produced by plotting

$$y(t) = \operatorname{sech}^2(t), \tag{1.1.17}$$

where sech= $1/\cosh$. An oscillatory soliton is depicted in Fig. 1.6 which is represented by

$$y(t) = \operatorname{sech}(t) \cdot \cos(t). \tag{1.1.18}$$

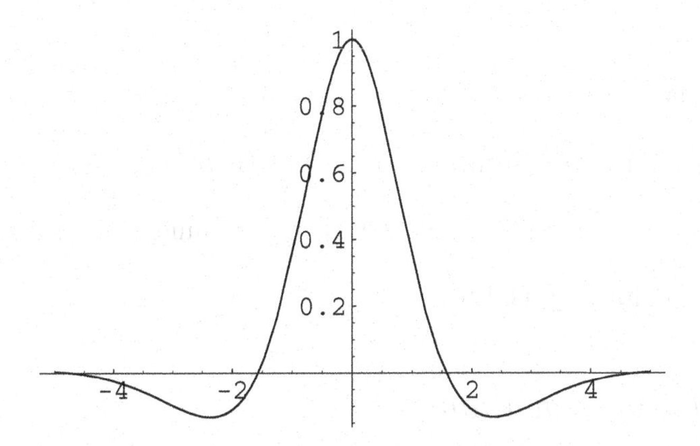


Fig. 1.6. Oscillatory soliton

The propagation of a harmonic wave (1.1.5) in space and time is shown in Fig. 1.7 by using the command

```
Plot3D[Cos[t-x],{x,-3*Pi,3*Pi},{t,0,3*Pi},Mesh->False,
PlotPoints->60,ColorOutput->GrayLevel]                    (1.1.19)
```

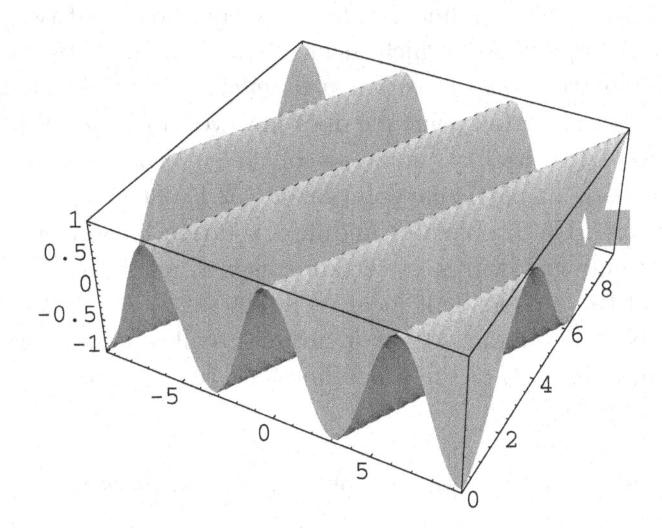


Fig. 1.7. Propagation of a harmonic wave in space and time

**Problems**

1. Verify the solution (1.1.2) of the LAPLACE equation (1.1.1).
2. Solve (1.1.1) by the setup $u(x, y, z, t) = f(x)g(y)h(z)j(t)$.
3. Verify the solution (1.1.3) of (1.1.1) in two dimensions $x, t$.
4. Derive the formula (1.1.9).

## *1.2 Linear wave equations*

The most general linear wave equation in two variables for a wave function $u(x, t)$ reads:

$$\begin{aligned} L(u) \equiv a(x,t)u_{xx} + 2b(x,t)u_{xt} + \\ c(x,t)y_{tt} + d(x,t)u_x + e(x,t)u_t + g(x,t)u = h(x,t). \end{aligned} \tag{1.2.1}$$

The coefficient functions $a, b, \ldots g$ describe the properties of the wave medium and the inhomogeneous term $h(x, t)$ describes exterior influences like forces. It is a general property of all linear differential equations that two particular solutions can be superposed, which means that the sum of two solutions is again a (new) solution (*superposition principle*). For nonlinear differential equations the superposition principle does however not hold. When writing down equation (1.1.6) we have made use of the superposition principle.

In order to obtain a particular solution of (1.2.1) one needs boundary and initial conditions. The problem of finding a solution to a given partial differential equation which will meet certain specified requirements for a given set of values of the independent local variables ($x_i, y_i, z_i$ - called boundary points) describing a boundary curve or surface is called a boundary problem. Since values $u(x_i, y_i, z_i, t)$ of the wave function are given, one uses the term *boundary value problem.*

Three types of boundary value problems are considered.

1. first boundary value problem (DIRICHLET *problem)*: Given a domain $R$ and its boundary surface $S$ and a function $f$ defined and continuous over $S$, then the DIRICHLET condition reads $u = f$ on the boundary where $u$ satisfies the three-dimensional wave equation for $u(x, y, z, t)$ and $u(x_i, y_i, z_i, t) = f(x_i, y_i, z_i, t)$.

2. second boundary value problem (NEUMANN *problem).* Here the normal derivative of the function $f$ is given on the boundary,

3. third boundary value problem: a linear combination

$$\frac{\partial u}{\partial n} + pu = m$$

is given.

Furthermore, an *initial condition* (CAUCHY *problem)* is important for wave equations. This condition fixes both value and normal derivative at exactly the *same* place (or time).

If the value $u$ or its derivatives or the function $m$ vanishes, the boundary value problem is called *homogeneous.* If the inhomogeneous term $h$ in (1.2.1) vanishes, the wave equation is termed homogeneous. If the differential equation or the boundary condition or both are inhomogeneous (they do not vanish), then the boundary problem is said to be *inhomogeneous.* Boundary curves or surfaces may be open or closed. A *closed boundary* surface (or curve) is one surrounding the boundary domain everywhere, confining it to a finite surface or volume. An open surface or curve does not completely enclose the domain, but lets it extend to infinity in at least one direction (*open boundary).*

Now we would like to classify linear partial differential equations of type (1.2.1). If the coefficient functions $a, b, c$ are constant but not equal ($a \neq b$), then the medium is *anisotropic.* If $a, b, c$ depend on location the medium is *inhomogeneous.* Since the inhomogeneous term $h$ in (1.2.1) describes external influences, it is sufficient to investigate the homogeneous equation. We introduce the following abbreviation

$$u_x = p, \quad u_t = q, \quad u_{xx} = r, \quad u_{xt} = s, \quad u_{tt} = v, \tag{1.2.2}$$

so that (1.2.1) reads

$$ar + 2bs + cv = F(u, p, q, x, t). \tag{1.2.3}$$

We now are interested in the question if a solution of (1.2.3) exists that satisfies the given CAUCHY (initial) conditions on the boundary: both $u$ and its normal derivative $\partial u/\partial n$ are prescribed. Since $u$ is given, then so is $\partial u/\partial s$. From $\partial u/\partial s$ and $\partial u/\partial n$ one can calculate $u_x = p$ and $u_t = q$. Thus the following relations are valid in general and on the boundary

$$\begin{aligned} dp &= du_x = r dx + s dt = u_{xx} dx + u_{xt} dt \\ dq &= du_t = s dx + v dt = u_{xt} dx + u_{tt} dt. \end{aligned} \tag{1.2.4}$$

These two expressions together with (1.2.3) constitute three linear equations for the determination of the three variables $r, s, v$ on the boundary. The determinant of this linear systems reads

$$\begin{vmatrix} a & 2b & c \\ \mathrm{d}x & \mathrm{d}t & 0 \\ 0 & \mathrm{d}x & \mathrm{d}t \end{vmatrix} = a\mathrm{d}t^2 - 2b\mathrm{d}x\mathrm{d}t + c\mathrm{d}x^2. \tag{1.2.5}$$

Only when this determinant vanishes, the system of equations has a solution. This gives

$$\left(\frac{\mathrm{d}x}{\mathrm{d}t}\right)^2 - \frac{2b}{c}\frac{\mathrm{d}x}{\mathrm{d}t} + \frac{a}{c} = 0 \tag{1.2.6}$$

and the solution is

$$\frac{\mathrm{d}x}{\mathrm{d}t} = \frac{b}{c} \pm \frac{1}{c}\sqrt{b^2 - ac}. \tag{1.2.7}$$

It clearly depends on the three functions $a, b, c$ if one obtains one real, two distinct real or two conjugate complex expressions for the curves $x(t)$ or $t(x)$ which are called MONGE *characteristics.*

1. If $b^2 - ac > 0$, then the curves $x(t)$ form two distinct families and the partial differential equation (1.2.1) is called *hyperbolic,*

2. if $b^2 - ac < 0$, then the characteristics are conjugate complex and (1.2.1) is called elliptic,

3. if $b^2 - ac = 0$ we have the parabolic type and only one real family of characteristics exists.

Each of the three equations can be transformed into a normal form. Introducing new coordinates called *characteristics*

$$\xi = \varphi(x,t), \quad \eta = \psi(x,t) \tag{1.2.8}$$

one obtains

$$\begin{aligned}
u_x &= u_\xi \varphi_x + u_\eta \psi_x, \quad u_t = u_\xi \varphi_t + u_\eta \psi_t \\
u_{xx} &= u_{\xi\xi} \varphi_x^2 + 2u_{\xi\eta} \varphi_x \psi_x + u_{\eta\eta} \psi_x^2 \dots \\
u_{xt} &= u_{\xi\xi} \varphi_x \varphi_t + u_{\xi\eta}(\varphi_x \psi_t + \varphi_t \psi_x) + u_{\eta\eta} \psi_x \psi_t \dots \\
u_{tt} &= u_{\xi\xi} \varphi_t^2 + 2u_{\xi\eta} \varphi_t \psi_t + u_{\eta\eta} \psi_t^2 \dots,
\end{aligned}$$

where the dots only represent first derivatives. Insertion into (1.2.3) and use of

$$a\varphi_x^2 + 2b\varphi_x\varphi_t + c\varphi_t^2 = 0 \quad (1.2.9)$$

(valid also for $\psi$) yields the normal form (1.2.10). In the calculation the coefficients of $u_{\xi\xi}$ and $u_{\eta\eta}$ had vanished due to $\varphi = \text{const}$ and $\varphi_x dx + \varphi_t dt = 0$. Then the *normal form* of the hyperbolic type reads

$$u_{\xi\eta} = F(u, u_\xi, u_\eta, \xi, \eta). \quad (1.2.10)$$

For the normal form of the elliptic type one receives

$$u_{\xi\xi} + u_{\eta\eta} = F(u, u_\xi, u_\eta, \xi, \eta) \quad (1.2.11)$$

when the transformation $\xi + i\eta = \varphi(x,t)$, $\xi - i\eta = \psi(x,t)$ is used. The parabolic case is not of interest for tsunamis or hurricanes.

It is now possible to prove [1.2] the *solvability* of boundary value problems. We summarize the results in Table 1.1.

Table 1.1. Solvability

| Boundary condition | Equation: hyperbolic | elliptic | parabolic |
|---|---|---|---|
| CAUCHY | | | |
| open boundary | *solvable* | indeterminate | overdeterminate |
| one closed boundary | overdeterminate | overdeterminate | overdeterminate |
| DIRICHLET | | | |
| open boundary | indeterminate | indeterminate | *solvable* |
| one closed boundary | indeterminate | *solvable* | overdeterminate |
| NEUMANN | | | |
| open boundary | indeterminate | indeterminate | *solvable* |
| one closed boundary | indeterminate | *solvable* | overdeterminate |

In this connection the term solvable means solvable by an analytic function.

## Problems

1. Show that $u_{xx} + u_{yy} = 0$ is an elliptic equation.

2. Show that $u_{xx} - u_{tt} = h(x,t)$ is a hyperbolic equation.

3. Investigate the type of the TRICOMI *equation*

$$u_{xx} + xu_{tt} = 0. \tag{1.2.12}$$

4. Derive the characteristics of (1.2.12).
Solution:

$$t \pm \frac{2}{3}(-x)^{3/2} = \text{const for } x < 0 \text{ and } t + \frac{2}{3}(ix)^{3/2} = \text{const for } x > 0. \tag{1.2.13}$$

5. Find the type of the EULER *equation* ($a, b, c$ are constants.)

$$au_{xx} + 2bu_{xy} + cu_{yy} = 0. \tag{1.2.14}$$

6. Find the type of $u_{xx} + 2u_{xy} + u_{yy} + x = 0$ (parabolic).
7. Investigate $u_{tt} = x^2 u_{xx} + u/4$. ($u(x,t) = \sqrt{x} f(\ln x - t)$).
8. Prove that the *superposition principle* is valid for equations of type (1.2.1).

## *1.3 Solutions of linear wave equations*

Elliptic differential equations are not of interest in the discussion of tsunamis, but play a certain role for hurricanes. We just want to mention that there are some similarities to hyperbolic equations, see (1.1.1),

$$\begin{aligned} u(x,t) &= f(x+ct) + g(x-ct) \quad \text{solves} \quad c^2 u_{xx} = u_{tt} \quad \text{and} \\ u(x,y) &= f(x+iy) + g(x-iy) \quad \text{solves} \quad u_{xx} + u_{yy} = 0. \end{aligned}$$

We now discuss solutions of linear partial differential equations of second order of *hyperbolic* type. The *normal form* (1.2.10) of the most general homogeneous linear wave equation (1.2.1) can be obtained formally by setting $h = 0$, $a = 0$, $c = 0$, $b = 1/2$:

$$L \equiv u_{xt} + du_x + eu_t + gu = 0. \tag{1.3.1}$$

In order to be able to proceed we need a short mathematical excursion. Let us consider two differential operators $L$ like (1.2.1), (1.3.1) and another, called *adjoint operator* $M(v)$ defined by

$$P \equiv vL(u) - uM(v) = \frac{\partial X}{\partial x} + \frac{\partial Y}{\partial t}, \tag{1.3.2}$$

where $X$ and $Y$ are functions of $v$ and $u$. Here $M$ is defined by the requirement that the expression $P$ be integrable and may be a kind of a divergence of the pseudo-vector with the components $X$ and $Y$. The problem is now to find $M, X, Y$. As soon as $v$ is known, we also know $u$. Using several identities like

$$\begin{aligned} avu_{xx} - u(av)_{xx} &= \frac{\partial}{\partial x}\left(avu_x - u\frac{\partial av}{\partial x}\right), \\ bvu_{xt} - u(bv)_{xt} &= \frac{\partial}{\partial x}(bvu_t) - \frac{\partial}{\partial t}\left(u\frac{\partial bv}{\partial x}\right), \\ dvu_x - u\frac{\partial}{\partial x}(-dv) &= \frac{\partial}{\partial x}(dvu) \quad \text{etc} \end{aligned}$$

one gets

$$M(v) = \frac{\partial^2 av}{\partial x^2} + 2\frac{\partial^2 bv}{\partial x \partial t} + \frac{\partial^2 cv}{\partial t^2} - \frac{\partial dv}{\partial x} - \frac{\partial ev}{\partial t} + gv = 0, \tag{1.3.3}$$

$$\begin{aligned} X &= a(vu_x - uv_x) + b(vu_t - uv_t) + (d - a_x - b_t)uv, \\ Y &= b(vu_x - uv_x) + c(vu_t - uv_t) + (e - b_x - c_t)uv. \end{aligned}$$

$M$ is the adjoint operator for $L$ which we wanted to find. The expressions $X$ and $Y$ are known as soon as $v$ is found. The special case $L(u) = M(v)$ is called self-adjointness. $L$ and $M$ are called *self-adjoint* (and exhibit special important properties in many other fields of physics). The condition of self-adjointness may now be written in the form

$$a_x + b_t = d; \quad b_x + c_t = e. \tag{1.3.4}$$

With these concepts at hand we may start the integration of linear hyperbolic partial differential equations [1.3]. In order to obtain the normal form (1.3.1), we again use $a = 0, c = 0, b = 1/2$ to obtain $M$ from (1.3.3) in the form

$$M(v) = v_{xt} - \frac{\partial dv}{\partial x} - \frac{\partial ev}{\partial t} + gv = 0, \tag{1.3.5}$$

as well as

$$\begin{aligned} X &= \frac{1}{2}(vu_t - uv_t) + duv \\ Y &= \frac{1}{2}(vu_x - uv_x) + euv. \end{aligned} \tag{1.3.6}$$

Inserting now $L$ from (1.3.1), $M$ from (1.3.5) and (1.3.6) into GAUSS *theorem* $\int \mathrm{div}\vec{A}df = \int A_n ds$ or

$$\int_S [vL(u) - uM(v)]\mathrm{d}f = \int \left( \frac{\partial X}{\partial x} + \frac{\partial Y}{\partial t} \right) \mathrm{d}f = \int_C [X\cos(n,x) + Y\cos(n,t)]\mathrm{d}s.$$

We get

$$\int_S [vL(u) - uM(v)]\mathrm{d}f = \int_C [X\cos(n,x) + Y\cos(n,t)]\mathrm{d}s. \tag{1.3.7}$$

Here $C$ is the boundary curve and $S$ is the region. According to Table 1.1 the hyperbolic equation is solvable for an open boundary. RIEMANN has therefore chosen the region $S$ as given in Fig. 1.8. This figure has been produced by the following Mathematica command:

```
Clear[l,G,r,Ci,GT0,GT1,GT2,GT3,GT4,GT5,GT6];
l=Line[{{1.,1.},{4.,1.},{4.,4.}}];
G=Graphics[l,Axes->True, AxesLabel->{''x'',''t''},
AspectRatio->1.];
r=3.; $DefaultFont={''Courier-Bold'',10};
Ci=Circle[{4.,1.},r];
G1=Graphics[Ci,AspectRatio->1.,
PlotRange->{{0,4.08},{0,4.08}}];
T0=Text[P,{4.095,1.}];
T1=Text[P1,{0.90,1.}];
T2=Text[P2,{4.,4.1}];
T3=Text[S,{3.,2.}];
T4=Text[C,{1.5,3.}];
T5=Text[eta,{-0.52,1.}];
T6=Text[xi,{1.,-0.37}];
GT0=Graphics[T0];
GT1=Graphics[T1];
GT2=Graphics[T2];
GT3=Graphics[T3];
GT4=Graphics[T4];
GT5=Graphics[T5];
GT6=Graphics[T6];
Show[G,G1,GT0,GT1,GT2,GT3,GT4,GT5,GT6,AspectRatio->1.,
PlotRange->{{-.9,4.5},{-0.6,4.5}}];
```

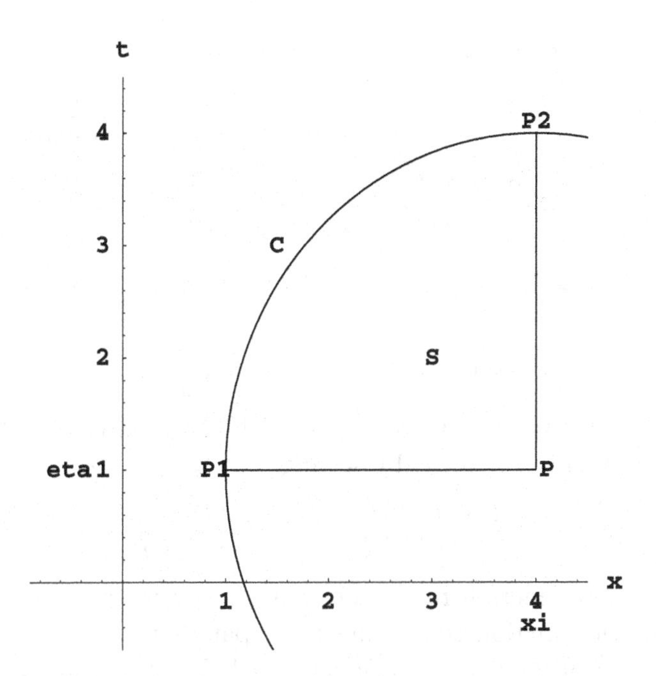


Fig. 1.8. RIEMANN integration (xi = $\xi$, eta = $\eta$)

To define the triangular region $S$ of Fig. 1.8 we first define two points $P_1(x_1, t_1)$ and $P_2(x_2, t_2)$ on the open (infinite) boundary curve $C$ which is situated in the $t(x)$ plane. Then we draw two straight lines parallel to the $x$- and $t$-axes, respectively. The lines start in $P_1$ and in $P_2$, respectively. They meet each other in $P(x = xi = \xi,\ t = \eta = \text{eta})$. The hyperbolic equation is only solvable for a CAUCHY *condition*. That is that the values and the normal derivative of $u$ on $C$ must be given, compare Table 1.1.

According to RIEMANN the following assumptions are made

1. $$M(v(x,t)) = 0 \quad \text{in } S, \tag{1.3.8}$$
2. $$v(P) = v(x = \xi, t = \eta) = 1, \quad (P \text{ is not on } C), \tag{1.3.9}$$
3. $$v_t - dv = 0 \quad \text{for} \quad x = \xi, v_x - ev = 0 \quad \text{for} \quad t = \eta. \tag{1.3.10}$$

These three conditions determine the function $v$ and its behavior along the characteristics $x = \xi$ and $t = \eta$. Apart from a factor, $v$ is RIEMANN's *hypergeometric function.*

Specializing now $d$, $e$ and $g$ in the anisotropic equation (1.3.5) for hydrodynamics, we have [1.3]

$$d = -\frac{\alpha}{x+t}, \quad e = -\frac{\alpha}{x+t}, \quad g = 0. \tag{1.3.11}$$

Then (1.3.10) yields for $v(x,t)$

$$\frac{1}{v}v_t = -\frac{\alpha}{x+t} = -\frac{\alpha}{\xi+t}, \quad \frac{1}{v}v_x = -\frac{\alpha}{x+t} = -\frac{\alpha}{x+\eta}. \tag{1.3.12}$$

Integration of (1.3.12) delivers

$$v = C_1(\xi+t)^{-\alpha}, \quad v = C_2(\xi+\eta)^{-\alpha}. \tag{1.3.13}$$

To obtain the solution of (1.3.5), (1.3.8) we take into account (1.3.9) which gives

$$C_1 = C_2 = (\xi+\eta)^{\alpha}, \quad v = \left(\frac{\xi+\eta}{x+t}\right)^{\alpha}, \tag{1.3.14}$$

so that $v(x=\xi, t=\eta) = 1$ according to (1.3.9) is satisfied. RIEMANN has modified the solution (1.3.14) by setting

$$v(x,t;\xi,\eta) = \left(\frac{\xi+\eta}{x+t}\right)^{\alpha} \cdot F(z), \quad z = -\frac{(x-\xi)(t-\eta)}{(x+t)(\xi+\eta)}. \tag{1.3.15}$$

Here $F$ is a new function to be determined. It will turn out that it is the hypergeometric function containing three parameters, all depending on $\alpha$. Taking into account $a=0, c=0, g=0, b=1/2$ and (1.3.11), the differential equation (1.3.3), (1.3.5) assumes now the form

$$M(v) = \frac{\partial^2 v}{\partial x \partial t} + \frac{\alpha}{x+t}\left(\frac{\partial v}{\partial x} + \frac{\partial v}{\partial t}\right) - 2\alpha\frac{v}{(x+t)^2} = 0. \tag{1.3.16}$$

This equation is satisfied by $v$ (1.3.14) but also by (1.3.15). We now will find $F$ [1.3]. Using (1.3.15) we obtain

$$\begin{aligned} v_x &= \left(\frac{\xi+\eta}{x+t}\right)^{\alpha}\left(\frac{-\alpha}{x+t}F(z) + F'(z)\frac{\partial z}{\partial x}\right), \\ v_t &= \left(\frac{\xi+\eta}{x+t}\right)^{\alpha}\left(\frac{-\alpha}{x+t}F(z) + F'(z)\frac{\partial z}{\partial y}\right). \end{aligned} \tag{1.3.17}$$

Insertion into (1.3.16) delivers

$$F''(z)\frac{\partial z}{\partial x}\frac{\partial z}{\partial t} + F'(z)\frac{\partial^2 z}{\partial x \partial t} - \frac{\alpha(\alpha+1)}{x+t}F(z) = 0. \tag{1.3.18}$$

Now the derivative of $z(x,t;\xi,\eta)$ can be expressed in the form

$$\frac{\partial z}{\partial x}\frac{\partial z}{\partial t} = \frac{1}{(x+t)^2}(z^2 - z),$$

$$\frac{\partial^2 z}{\partial x \partial t} = \frac{2z-1}{(x+t)^2}.$$

Using these expressions (1.3.18) becomes

$$z(1-z)F'' + (1-2z)F' + \alpha(\alpha+1)F = 0. \tag{1.3.19}$$

This equation defines a hypergeometric function. It is usual to define a more general hypergeometric function by the differential equation [1.1]

$$z(1-z)F'' + [\gamma - (\delta+\beta+1)z]F' - \delta\beta F = 0. \tag{1.3.20}$$

Comparison to (1.3.10) yields $\gamma = 1, \delta = 1 - \beta = 1 + \alpha, \beta = -\alpha$. Since the solution of (1.3.20) is usually designated by $F(z; \delta, \beta, \gamma)$ one has the solution of (1.3.19) in the form

$$F(z; 1+\alpha, -\alpha, 1). \tag{1.3.21}$$

Series representation, the Mathematica expressions

```
DSolve[y''[x]+(c-(a+b+1)*x)*y'[x]/(x*(1-x))
-a*b*y[x]/(x*(1-x))==0,  y[x],x]
y[x]=C[1]*Hypergeometric2F1[a,b,c,x]+(-1)^(1-c)*x^(1-c)+
C[2]*Hypergeometric2F1[1+a-c,1+b-c,2-c,x]                (1.3.22)
```

and details on the hypergeometric function family (GAUSS function) may be found in the specialized mathematical literature [1.1], [1.2], [1.4], [1.5].

We now have presented some tools to solve linear hyperbolic partial differential equations of second order with variable (and constant) coefficients. These tools will be used in some sections of chapter 3. Tsunamis are however nonlinear waves and when discussing nonlinear wave equations we will find that some tools presented here will be useful.

To conclude this section we will solve a linear wave equation in an isotropic inhomogeneous medium. We choose the TRICOMI *equation* (1.2.12)

$$u_{xx} + xu_{tt} = 0. \tag{1.3.23}$$

From (1.2.1) we read $a = 1, b = 0, c = x$. Then (1.2.7) yields the differential equation for the characteristics

$$\frac{dx}{dt} = \pm\frac{1}{x}\sqrt{-x}. \tag{1.3.24}$$

The solutions are

$$t(x) = \text{const} \pm \frac{2}{3}(-x)^{3/2} \quad \text{for} \quad x < 0, \quad \text{hyperbolic}, \tag{1.3.25}$$

$$t(x) = \text{const} \pm \frac{2}{3}(ix)^{3/2} \quad \text{for} \quad x > 0, \quad \text{elliptic}. \tag{1.3.26}$$

A full solution of a boundary value problem shall be given for the following equation

$$u_{xx} + y^2 u_{yy} = 0 \quad \text{in the domain} \quad S\left\{\begin{array}{ccccc} 0 & \leq & x & \leq & a \\ 0<b & \leq & y & \leq & c \end{array}\right. \tag{1.3.27}$$

where $a, b, c$ are given constants. The boundary values should be

$$u(0,y) = u(a,y) = 0, \tag{1.3.28}$$

$$u(x,b) = f(x), \quad u(x,c) = g(x), \tag{1.3.29}$$

where $f(x)$ and $g(x)$ are given functions.

The characteristics indicate the elliptic type for $y \neq 0$ and the parabolic type for $y = 0$. Setting up

$$u(x,y) = \sum_{\nu=1}^{\infty} \varphi_\nu(x)\psi_\nu(y), \tag{1.3.30}$$

we try a direct solution. This must satisfy the boundary condition (1.3.28) $\varphi_\nu(0) = \varphi_\nu(a) = 0$. The ansatz $\varphi''_\nu(x) = -\lambda_\nu \varphi_\nu(x)$ yields

$$\varphi_\nu = \sin\frac{\nu\pi x}{a}, \quad \lambda_\nu = \frac{\nu^2\pi^2}{a^2}, \quad \nu = 1,2,3\dots. \tag{1.3.31}$$

If one inserts (1.3.30) into (1.3.27) one obtains

$$\begin{aligned} u_{xx} + y^2 u_{yy} &= \sum_{\nu=1}^{\infty} (\varphi''_\nu \psi_\nu + y^2 \varphi_\nu \psi''_\nu) = \\ &= \sum_{\nu=1}^{\infty} \sin\frac{\nu\pi x}{a}\left[y^2\psi''_\nu(y) - \frac{\nu^2\pi^2}{a^2}\psi_\nu(y)\right] = 0, \end{aligned} \tag{1.3.32}$$

so that

$$y^2\psi''_\nu - \frac{\nu^2\pi^2}{a^2}\psi_\nu(x) = 0, \quad \nu = 1,2,3\dots \tag{1.3.33}$$

has to be solved together with the boundary conditions (1.3.29)

$$\begin{aligned} \psi_\nu(b) &= \frac{2}{a}\int_0^a f(x)\sin\frac{\nu\pi x}{a}dx = b_\nu, \\ \psi_\nu(c) &= \frac{2}{a}\int_0^a g(x)\sin\frac{\nu\pi x}{a}dx = c_\nu. \end{aligned} \tag{1.3.34}$$

In order to solve this boundary value problem we make the setup

$$\psi_\nu(y) = e_\nu y^{\alpha_\nu} + f_\nu y^{-\beta_\nu}. \tag{1.3.35}$$

This expression satisfies the differential equation (1.3.33), if

$$\alpha_\nu = \frac{1}{2} + \sqrt{\frac{1}{4} + \frac{\nu^2\pi^2}{a^2}}, \quad \beta_\nu = -\frac{1}{2} + \sqrt{\frac{1}{4} + \frac{\nu^2\pi^2}{a^2}}. \tag{1.3.36}$$

The new constants $e_\nu$ and $f_\nu$ in (1.3.35) are then given from the boundary condition

$$\begin{aligned} e_\nu &= \frac{c_\nu c^{\beta_\nu} - b_\nu b^{\beta_\nu}}{c^{\alpha_\nu+\beta_\nu} - b^{\alpha_\nu+\beta_\nu}}, \\ f_\nu &= \frac{(bc)^{\beta_\nu}\left(b_\nu c^{\alpha_\nu} - c_\nu b^{\alpha_\nu}\right)}{c^{\alpha_\nu+\beta_\nu} - b^{\alpha_\nu+\beta_\nu}} \quad \text{for } b \neq c. \end{aligned} \tag{1.3.37}$$

Collecting all together one has the solution of (1.3.27) in the form

$$\begin{aligned} u(x,y) &= \sum_{\nu=1}^{\infty} \frac{\sin\dfrac{\nu\pi x}{a}}{c^{\alpha_\nu+\beta_\nu} - b^{\alpha_\nu+\beta_\nu}} \\ &\quad \cdot \left[\left(c_\nu c^{\beta_\nu} - b_\nu b^{\beta_\nu}\right) y^{\alpha_\nu} + (bc)^{\beta_\nu}\left(b_\nu c^{\alpha_\nu} - c_\nu b^{\alpha_\nu}\right) y^{-\beta_\nu}\right] \end{aligned} \tag{1.3.38}$$

for $0 < b < c$. If however $b = 0$, then the boundary condition $u(x,0) = f(x)$ can no longer be given! The solution of (1.3.27) is then only determined by $u(x,c) = g(x)$ and reads

$$u(x,y) = \sum_{\nu=1}^{\infty} c_\nu \sin\frac{\nu\pi x}{a} \cdot \left(\frac{y}{c}\right)^{\alpha_\nu}. \tag{1.3.39}$$

We thus see that even the solution of simple linear partial differential equations of second order is complicated, if the coefficients are variable.

## Problems

1. Using (1.3.11) verify the solution (1.3.14) of (1.3.5).

2. Derive (1.3.18).

## *1.4 Nonlinear wave equations*

In chapters 2 and 3, it will turn out that many types of waves in water or air have to be described by nonlinear partial differential equations of second order.

In hydrodynamics we will have the situation that one starts with a system of nonlinear partial differential equations of first order, see chapter 2. One then has to derive wave equations from these equations of first order. Therefore we will first discuss the characteristics of one and later of several nonlinear partial differential equations of first order. In order to make clear the terminology, we discuss some equations for $u(x,t)$.

$$\begin{aligned} u_x &= u_t && \text{first order linear,} \\ x^2 u_x &= u_t, u_x a(x,t) = u_t && \text{linear, but variable coefficients,} \\ u_x^2 &= u_t && \text{nonlinear,} \\ u u_x &= u_t && \text{quasilinear (derivatives are linear).} \end{aligned}$$

Fortunately, the hydrodynamic equations which we will discuss are quasilinear.

LAGRANGE has shown that the most general quasilinear partial differential equation of first order

$$P(x,y,u)u_x(x,y) + Q(x,y,u)u_y(x,y) = R(x,y,u), \tag{1.4.1}$$

where $P, Q, R$ are arbitrary in $x$ and $y$ but linear in $u$ possesses an implicit general solution in the form

$$F\left(\varphi(x,y,u), \psi(x,y,u)\right) = 0,$$

where $F$ is an arbitrary differentiable function and where $\varphi, \psi$ satisfy

$$\varphi(x,y,u) = \text{const} = a, \quad \psi(x,y,u) = \text{const} = b \tag{1.4.2}$$

and are two independent solutions of any combination of the differential equations of the characteristics

$$\frac{\mathrm{d}x}{P} = \frac{\mathrm{d}y}{Q} = \frac{\mathrm{d}u}{R} \quad \text{or} \quad \frac{\mathrm{d}x}{\mathrm{d}s} = P, \quad \frac{\mathrm{d}y}{\mathrm{d}s} = Q, \quad \frac{\mathrm{d}u}{\mathrm{d}s} = R. \tag{1.4.3}$$

Integration yields two surfaces (1.4.2). Their cut delivers curves in space. $s$ is the arc length along these curves. Let us consider an example $u(x,t)$

satisfying

$$u_t + uu_x = 0. \tag{1.4.4}$$

We read from (1.4.1) that $P = 1, Q = u, R = 0$ and thus

$$\frac{\mathrm{d}t}{\mathrm{d}s} = 1, \quad \frac{\mathrm{d}x}{\mathrm{d}s} = u, \quad \frac{\mathrm{d}u}{\mathrm{d}s} = 0. \tag{1.4.5}$$

Integration yields $t = s + c, \mathrm{d}t = \mathrm{d}s, u = a$. Due to

$$\frac{\mathrm{d}x}{\mathrm{d}s} = \frac{\mathrm{d}x}{\mathrm{d}t}\frac{\mathrm{d}t}{\mathrm{d}s} = \frac{\mathrm{d}x}{\mathrm{d}t} = u \tag{1.4.6}$$

one obtains $x = ut + b$ and the two characteristics $\varphi = a = u, \psi = b = x = ut$. Here $a, b$ are integration constants. The solution of (1.4.4) is then given by

$$F(a, b) = F\left(\varphi(x, t, u), \psi(x, t, u)\right) = F(u, x - ut) = 0. \tag{1.4.7}$$

Now we are able to solve the CAUCHY-problem

$$u(x, t = 0) = f(x). \tag{1.4.8}$$

Replacement ("enlargement") of the argument $x \to x - ut$ yields the solution $u(x, t) = f(x - ut)$. This expression satisfies the initial condition (1.4.8) and the equation (1.4.4)

For an implicit nonlinear partial differential equation of first order

$$F\left(u(x, y), u_x(x, y), u_y(x, y), x, y\right) = 0, \tag{1.4.9}$$

the procedure has to be modified and the derivatives $u_x$ and $u_y$ as well as $u$ itself have to be assumed as five independent variables. The characteristics are then determined by

$$\begin{aligned} &\frac{\mathrm{d}x}{F_{u_x}} = \frac{\mathrm{d}y}{F_{u_y}} = \frac{\mathrm{d}u}{u_x F_{u_x} + u_y F_{u_y}} \\ &= \frac{\mathrm{d}u_x}{-u_x F_u - F_x} = \frac{\mathrm{d}u_y}{-u_y F_u - F_y}. \end{aligned} \tag{1.4.10}$$

These five ordinary differential equations have to be solved. Let us consider an example. We choose the following nonlinear partial differential equation of first order

$$F \equiv 16u_x^2 u^2 + 9u_y^2 u^2 + 4u^2 - 4 = 0. \tag{1.4.11}$$

From (1.4.10) one gets

$$\begin{aligned} F_x = 0, \quad F_y = 0, \quad F_u = (32u_x^2 + 18u_y^2)u + 8u, \\ F_{u_x} = 32u_x u^2, \quad F_{u_y} = 18u_y u^2 \end{aligned} \tag{1.4.12}$$

and thus from (1.4.10)

$$\begin{aligned} \frac{\mathrm{d}x}{32u_x u^2} = \frac{\mathrm{d}y}{18u_y u^2} = \frac{\mathrm{d}u}{32u_x^2 u^2 + 18u_y^2 u^2} \\ = \frac{\mathrm{d}u_x}{-32u_x^3 u - 18u_y^2 u u_x - 8uu_x} = \frac{\mathrm{d}u_y}{-32u_x^2 u u_y - 18u_y^3 u - 8uu_y}. \end{aligned} \tag{1.4.13}$$

Multiplication of the five fractions by 1, 0, $4u_x$, $4u$ and 0, respectively, collects the five fractions over a common denominator $N$ into one fraction

$$\begin{aligned} N = \; & 1 \cdot (+32u_x u^2) + 4u_x \cdot (+32u_x^2 u^2 + 18u_y^2 u^2) \\ & -4u \cdot (32u_x^3 u + 18u_x u_y^2 u + 8u_x u) = 0 \end{aligned} \tag{1.4.14}$$

and hence

$$\mathrm{d}x + 4u_x \mathrm{d}u + 4u\mathrm{d}u_x = 0 \quad \text{or} \quad \mathrm{d}x + 4\mathrm{d}(u_x u) = 0. \tag{1.4.15}$$

Integration and inserting of $u_x$ yields

$$(x-a)^2 + 9u_y^2 u + 4u^2 - 4 = 0. \tag{1.4.16}$$

$a$ is the integration constant. $u_x$ has been calculated from (1.4.11). After elimination of $u_y$ one obtains the solution of (1.4.11) in the implicit form

$$\frac{(x-a)^2}{4} + \frac{4(y-b)^2}{9} + u^2 - 1 = 0. \tag{1.4.17}$$

The basic hydrodynamic equations have the form of several quasilinear partial differential equations of first order

$$\begin{aligned} a_{11}u_x + a_{12}v_x + b_{11}u_y + b_{12}v_y = h_1(x,y), \\ a_{21}u_x + a_{22}v_x + b_{21}u_y + b_{22}v_y = h_2(x,y). \end{aligned} \tag{1.4.18}$$

Here $u(x,y)$, $v(x,y)$ are two dependent variables and the coefficients $a_{ik}(x,y,u,v)$, $b_{ik}(x,y,u,v)$ are not constant, but only linear functions of

$u$ and $v$. Introducing the two vectors $\vec{u} = \{u, v\}$, $\vec{h} = \{h_1, h_2\}$ and the matrices $A\{a_{ik}\}$ and $B\{b_{ik}\}$ one may rewrite (1.4.18) in the form

$$A\vec{u}_x + B\vec{u}_y = \vec{h}. \tag{1.4.19}$$

Now we look for the characteristics of this system. They would probably be of the form $\psi(x, y) = \text{const}$ or possibly $y = k(x) + \text{const}$, $\mathrm{d}y/\mathrm{d}x = k'(x)$. Then the following two equations are valid along the characteristic curve

$$\begin{aligned} \frac{\mathrm{d}v}{\mathrm{d}x} &= \frac{\partial v}{\partial x} + \frac{\partial v}{\partial y} \cdot \frac{\mathrm{d}y}{\mathrm{d}x} = v_x + v_y k', \\ \frac{\mathrm{d}u}{\mathrm{d}x} &= \frac{\partial u}{\partial x} + \frac{\partial u}{\partial y} \cdot \frac{\mathrm{d}y}{\mathrm{d}x} = u_x + u_y k'. \end{aligned} \tag{1.4.20}$$

Mathematica may help to obtain $u_x$ and $v_x$. Introducing the notation $u_x \to ux$, $v_x \to vx$, $\mathrm{d}v/\mathrm{d}x \to dvx$, $\mathrm{d}u/\mathrm{d}x \to dux$, $k' \to ks$ we may use

```
Solve[{vx+vy*ks==dvx,ux+uy*ks==dux},{ux,vx}]        (1.4.21)
```

to solve (1.4.20) for $v_x$, $u_x$. Inserting the solutions $v_x = \mathrm{d}v_x/\mathrm{d}x - v_y k'$ and $u_x = \mathrm{d}u_x/\mathrm{d}x - u_y k'$ into (1.4.18) one obtains

$$\begin{aligned} u_y(-a_{11}k' + b_{11}) + v_y(-a_{12}k' + b_{12}) &= h_1 - a_{11}\frac{\mathrm{d}u}{\mathrm{d}x} - a_{12}\frac{\mathrm{d}v}{\mathrm{d}x}, \\ u_y(-a_{21}k' + b_{21}) + v_y(-a_{22}k' + b_{22}) &= h_2 - a_{21}\frac{\mathrm{d}u}{\mathrm{d}x} - a_{22}\frac{\mathrm{d}v}{\mathrm{d}x}. \end{aligned} \tag{1.4.22}$$

If the values of $u$ and $v$ are given along the characteristic curves $y(x)$, then the system (1.4.22) allows the calculation of $u_x$ and $v_y$ according to the CRAMER *rule*. We define

$$R = \begin{vmatrix} a_{11}k' - b_{11} & a_{12}k' - b_{12}, \\ a_{21}k' - b_{21} & a_{22}k' - b_{22} \end{vmatrix} = |Ay' - B|, \tag{1.4.23}$$

$$V_1 = \begin{vmatrix} a_{11}k' - b_{11} & h_1 - a_{11}\mathrm{d}u/\mathrm{d}x - a_{12}\mathrm{d}v/\mathrm{d}x \\ a_{21}k' - b_{21} & h_2 - a_{21}\mathrm{d}u/\mathrm{d}x - a_{22}\mathrm{d}v/\mathrm{d}x \end{vmatrix}, \tag{1.4.24}$$

$$V_2 = \begin{vmatrix} a_{12}k' - b_{12} & h_1 - a_{11}\mathrm{d}u/\mathrm{d}x - a_{12}\mathrm{d}v/\mathrm{d}x \\ a_{22}k' - b_{22} & h_2 - a_{21}\mathrm{d}u/\mathrm{d}x - a_{22}\mathrm{d}v/\mathrm{d}x \end{vmatrix}. \tag{1.4.25}$$

Three cases are possible:

1. $R \neq 0$: one can calculate $u_y, v_y$ and all first derivatives $u_k, v_k$ are determined along the characteristic curves,
2. $R = 0, V_1$ or $V_2 = 0$, the linear equations for $u_y, v_y$ are linearly depending and an infinite manifold of solutions $u_y, v_y$ exists,
3. $R = 0, V_1 \neq 0$ or $V_2 \neq 0$: no solutions $u_y, v_y$ exist.

Characteristics may also be derived for a system of quasilinear partial differential equations of second order. For $m$ independent variables $x_k$, $k = 1 \ldots m$, $l = 1 \ldots m$ and $n$ depending variables $u_j$, $j = 1 \ldots n$, such a system may be written as

$$\sum_{j=1}^{n} \sum_{k,l=1}^{m} A_{ij}^{(kl)}(x_k, u_j) \frac{\partial^2 u_j}{\partial x_k \partial x_l} + H\left(x_k, u_j, \frac{\partial u_j}{\partial x_l}\right) = 0, \quad i = 1 \ldots m. \tag{1.4.26}$$

The characteristics of "second order" of this system obey a partial differential equation of first order and of degree $n^2$, which reads

$$\left| \sum_{k,l=1}^{m} A_{ij}^{(kl)} \frac{\partial \varphi}{\partial x_k} \frac{\partial \varphi}{\partial x_l} \right| = 0. \tag{1.4.27}$$

This system has again to be solved using the characteristics of partial differential equations of first order.

Finally, we consider nonlinear partial differential equations of second order. If such equation is implicit and of the form

$$F(x, y, z, p, q, r, s, t) = 0, \tag{1.4.28}$$

where we used again the notation $u_x = p, u_y = q, u_{xx} = r, u_{xy} = s, u_{yy} = t$, then one can define

$$a = \frac{\partial F}{\partial z_{xx}}, \quad 2b = -\frac{\partial F}{\partial z_{xy}}, \quad c = \frac{\partial F}{\partial z_{yy}} \tag{1.4.29}$$

and insert into an equation $\mathrm{d}y/\mathrm{d}x$, analogous to (1.2.7). One then may classify the equation according to its type. The physics of waves described by nonlinear (quasilinear) wave equations will be discussed in the next section.

**Problems**

1. Verify the solution (1.4.17) of (1.4.11).

2. Solve

$$a\psi_{xx} + c\psi_{yy} + \psi_x = 0 \tag{1.4.30}$$

using the setup $\psi_x = u, \psi_y = v$ or $\psi_x = u, \psi_x + \psi_y = v$. Solve the resulting equations for $u(x,y), v(x,y)$.

## *1.5 Physics of nonlinear wave equations*

To understand the physics behind some mathematical terms in a wave equation we consider a weakly nonlinear wave equation [1.6] of the form

$$\frac{1}{c^2}\Phi_{tt} - \Phi_{xx} + b\,\Phi + \varepsilon g\,\Phi_t = -V'(\Phi) + \varepsilon N(\Phi_t) + \varepsilon\,\Phi_t G(\Phi). \tag{1.5.1}$$

This equation exhibits *frequency dispersion* $\omega(k)$, *amplitude dispersion* $\omega(k, A)$, *dissipation*, *nonlinear effects* and *modulation* of the amplitude and the phase of a wave. Here $V' = \mathrm{d}V/\mathrm{d}\Phi$, $V(\Phi)$, $G(\Phi)$ and the nonlinear dissipation $N = -\Phi_t^{2n-1}$ are given nonlinear functions, $c, b$ and $g$ are constants which may depend on the frequency $\omega = 2\pi\nu$. As usual, $\varepsilon$ is a small parameter. We now define a *phase surface*:

$$\Theta(x,t) = \text{const.} \tag{1.5.2}$$

This surface has the property that all points $(x,t)$ on it have the same value of the wave function $\Phi(x,t)$. We thus have

$$\mathrm{d}\Theta = \Theta_x \mathrm{d}x + \Theta_t \mathrm{d}t = 0, \tag{1.5.3}$$

so that points moving with the speed

$$\frac{\mathrm{d}x}{\mathrm{d}t} = -\frac{\Theta_t}{\Theta_x}, \tag{1.5.4}$$

see a constant phase $\Theta$. Defining now a *wave number* $k$ and a *frequency* $\omega$ by

$$k = \Theta_x, \quad \omega = -\Theta_t = \quad \text{or} \quad \frac{\partial\Theta}{\partial t} + \omega = 0, \tag{1.5.5}$$

we find that (1.5.4) defines the *phase speed.* In three-dimensional notation we have $\nabla\Theta = \vec{k}$, which defines the *wave vector* satisfying

$$\operatorname{curl}\vec{k} = 0, \tag{1.5.6}$$

which indicates that *wave crests* are neither vanishing nor splitting off. The last two equations result in the conservation of wave crests

$$\frac{\partial \vec{k}}{\partial t} + \nabla\omega = 0. \tag{1.5.7}$$

A point moving with the *group velocity* (1.1.12)

$$c_g = \frac{\mathrm{d}\omega}{\mathrm{d}k} \tag{1.5.8}$$

sees $\omega$ unchanged.

Looking up the different terms in (1.5.1), we may dress up the following classification:

1. $N = 0,\ G = 0,\ V' = 0$: the equation is linear, $k$ and $\omega$ are independent of $x$ und $t$.

   1.1 $b = 0,\ g = 0$: no dispersion $\omega(k)$, no dissipation (no damping effects). The solution is:

   $$\Phi(x,t) = A\exp(ikx - i\omega t), \quad \omega = ck, \tag{1.5.9}$$

   see (1.1.6), (1.1.4).

   1.2 $b \neq 0,\ g = 0$, frequency dispersion $\omega(k)$, no dissipation, solution and *dispersion relation:*

   $$\Phi(x,t) = A\exp(ikx - i\omega t), \quad \omega(k) = \pm c\sqrt{k^2 + b^2}. \tag{1.5.10}$$

   1.3 $b = 0,\ g \neq 0$, dissipation, $\omega$ becomes complex

   $$\Phi(x,t) = A\exp(ikx - i\omega t), \quad \omega = \frac{1}{2}ic^2\varepsilon g \pm c\sqrt{k^2 - c^2\varepsilon^2 g^2/4}. \tag{1.5.11}$$

   1.4 $b \neq 0,\ g \neq 0$, dispersion and dissipation,

   $$\Phi(x,t) = A\exp(ikx - i\omega t), \quad D(\omega,k) = \omega^2/c^2 - k^2 - b + i\omega\varepsilon g = 0. \tag{1.5.12}$$

2. $N \neq 0,\ G \neq 0,\ V' \neq 0$: the wave equation is nonlinear and dissipative. $V'$ describes a strong nonlinearity, the terms containing $\Phi_t$ are dissipative, $\varepsilon N(\Phi_t)$ is a weak nonlinear dissipative term and the term $\varepsilon\Phi_t(\Phi)G$ describes weak dissipation together with strong nonlinearity.

Even for a weak nonlinearity one has frequency dispersion $\omega(k)$ and amplitude dispersion $\omega(A, k)$. Using stretched variables $X = \varepsilon x$, $T = \varepsilon t$ and an adapted KRYLOV-BOGOLYUBOV method two theorems can be derived [1.6]:

1. For a nonlinear conservative (nondissipative) wave equation of the type (1.5.1) the amplitude $A$ is constant and not modulated. Then the phase is given by

$$\Theta(x,t) = k(x,t)x - \omega(x,t)t, \tag{1.5.13}$$

see (1.5.5).

2. For any nonlinear dissipative wave equation of the type (1.5.1) the frequency $\omega$ is however not modified by the dissipation terms in first order of $\varepsilon$.

Furthermore it can be shown that a *stability* theorem holds. Let $\omega_0 = f(k)$ be the dispersion relation of the linear equation according to (1.5.12) and derive the quasilinear equation

$$\Theta_{tt} + 2f'\Theta_{xt} + (f'^2 + f''t)\Theta_{xx} = Q \tag{1.5.14}$$

for the phase $\Theta(x,t)$, $\Theta_x = k(x,t)$, then the effect of nonlinear terms on stability is described by $f''(k) \cdot [\omega - f(k)]$. The stability behavior of the linear equation, described by $\omega_0 = f(k)$ is not altered by nonlinear terms, if $f''(k) \cdot [\omega - f(k)] > 0$. If, however, $f''(k) \cdot [\omega - f(k)] < 0$, then the nonlinear terms may destabilize an otherwise stable solution of a linear equation. On the other hand the inclusion of a dissipative term $Q$ does not by itself modify the character of the stability behavior, but the time behavior of unstable and stable modes is modified.

Now we investigate a *modulated wave*. In first approximation we write down a sinusoidal wave

$$A_0 \cos\Theta_0, \quad \Theta_0 = k_0 x - \omega_0 t, \tag{1.5.15}$$

where $A_0$, $\Theta_0$, $k_0$ and $\omega_0$ are constants. Then we assume a slow amplitude variation $A(x,t)$ and a phase variation $\Theta(x,t)$

$$\Theta(x,t) = k_0 x - \omega_0 t + \varphi(x,t) \tag{1.5.16}$$

According to (1.5.5) we redefine

$$\omega(x,t) = -\Theta_t = \omega_0 - \varphi_t, \quad k(x,t) = \Theta_x = k_0 + \varphi_0. \tag{1.5.17}$$

For weak modulation one may expand [1.7]

$$\omega = \omega_0 + \frac{\partial \omega}{\partial A_0^2}\left(A^2 - A_0^2\right) + \frac{\partial \omega}{\partial k_0}(k - k_0) + \frac{\partial^2 \omega}{\partial k_0^2}(k - k_0)^2 + \dots . \tag{1.5.18}$$

Making the replacement

$$\omega - \omega_0 \rightarrow i\frac{\partial}{\partial t}, \quad k - k_0 \rightarrow i\frac{\partial}{\partial x} \tag{1.5.19}$$

one obtains the so-called *nonlinear* SCHROEDINGER *equation*

$$i\left[\frac{\partial A}{\partial t} + \frac{\partial \omega}{\partial k_0}\frac{\partial A}{\partial x}\right] + \frac{1}{2}\frac{\partial^2 \omega}{\partial k_0^2}\frac{\partial^2 A}{\partial x^2} - \frac{\partial \omega}{\partial A_0^2}|A|^2 A = 0. \tag{1.5.20}$$

This equation has nothing to do with the quantumtheoretical SCHROEDINGER equation.

In a frame of reference $\xi, \tau$ moving with the group velocity, equation (1.5.20) becomes [1.7]

$$i\frac{\partial A}{\partial \tau} + \frac{1}{2}\frac{\partial^2 A}{\partial \xi^2} + \alpha |A|^2 A = 0, \quad \alpha = -\frac{\partial \omega / \partial A_0^2}{\partial^2 \omega / \partial k_0^2}. \tag{1.5.21}$$

If one inserts the setup

$$A(\xi, \tau) = U(\xi - c\tau)\exp(ik\xi - i\omega\tau), \quad |A|^2 = U^2 \tag{1.5.22}$$

into (1.5.21) one gets as the real part

$$U'' + U(2\omega - k^2) + 2\alpha |U|^2 U = 0. \tag{1.5.23}$$

Multiplication of (1.5.23) by $U'$ and two integrations yield

$$\xi - c\tau = \int \frac{\mathrm{d}U}{\sqrt{(2\omega - k^2)U^2 - \alpha U^4 + C_1}} + C_2. \tag{1.5.24}$$

The integral in (1.5.24) is an *elliptic integral* and $U(\xi - c\tau)$ becomes a JACOBI *elliptic function*, see Fig. 1.2. The wave $U(\tau)$ is called a *cnoidal wave*. For $C_1 = 0$ one obtains an *envelope soliton*

$$U(\xi - c\tau) = \text{const} \cdot \text{sech}\left[(2\omega - k^2)(\xi - c\tau)\right]. \tag{1.5.25}$$

If one considers the real part of the solution of the nonlinear SCHROEDINGER equation (1.5.21), one gets [1.1]

$$A(\xi, \tau) = \text{const} \cdot \text{sech}\left[(2\omega - k^2)(\xi - c\tau)\right]\cos(k\xi - \omega\tau) \tag{1.5.26}$$

which describes an *oscillatory soliton*, see Fig. 1.6. The function sech (secans hyperbolicus) is defined by

$$\int \frac{\mathrm{d}x}{x\sqrt{1-x^2}} = -\operatorname{arsech} x. \tag{1.5.27}$$

$A(\xi, \tau)$ describes *amplitude modulation*.

# 2. Basic flow equations

## *2.1 Units and properties of substances*

When deriving flow equations one has to have in mind that *tsunamis* and *hurricanes* are quite intricate phenomena with an interplay of three media: water, air and vapor (steam). Three factors play important roles. There are first the properties of the media like *density* , *surface tension*, *compressibility*, *specific heat* etc, which enter into the differential equations. Second there are exterior factors like *Earth's rotation*, *gravity*, CORIOLIS *force* etc, which influence the motion and may enter into the equations. Third there are boundary and initial conditions like *pressure of wind*, *earthquakes* or free surfaces and the *depth* of the ocean or of a lake. Finally, thermodynamic considerations like *condensation*, condensation nuclei, diameter of droplets in vapor, *evaporation* and *vaporization* enter into the deliberations.

To describe all these factors one needs *units*. We will use the rules of the International Union of Pure and Applied Physics (SI-system) [2.1]. Later on we shall give some conversion factors for other (UK, US) units. The most important units and properties to be considered are: (s = seconds, g = gram, m = meters).

| | | | |
|---|---|---|---|
| Forces, Weight: | Newton N | $\mathrm{N = kg\ m\ s^{-2}}$ | |
| | or Kilopond | kp = 9.80665 N | and $\mathrm{dyn = 10^{-5} N}$ |
| Energy, Heat | Joule, J | J = N m | $\mathrm{kg\,m^2\,s^{-2}}$ |
| | or $\mathrm{erg = 10^{-7} J}$ | and cal = 4.1868 J | |
| Power | Watt, W | $\mathrm{W = J\,s^{-1}}$ | $\mathrm{10^7 erg\ s^{-1}}$ |
| Mass | Kilogramm kg, | also g | and mol |
| Density | $\rho$ | $\mathrm{kg\ m^{-3}}$, | $\mathrm{g\ cm^{-3}}$ |
| Pressure: | p Pascal, Pa | $\mathrm{Pa = N\,m^{-2}}$ | $\mathrm{J\ m^{-3}}$ |
| | | $\mathrm{= kg\ s^{-2} m^{-1}}$ | |
| | or Torr | or at (atmosphere) | and $\mathrm{bar = 10^5\ Pa}$ |
| | = 1.333 mb | at = 0.980665 bar | = 98.07 kPa |
| | 1 mbar = | 100 Pa = | 1 hPa |
| Temperature: | T, t in | (K = Kelvin degree, | |
| | °C, °K, °F, | absolute temperature) | |
| | T[° K] = 273.15 + t°C (Celsius), | | $\mathrm{t°C = \frac{5}{9}(t°F\text{-}32)}$. |

For the convenience of readers from the United Kingdom and the US we now give some conversion factors to other units [2.2]. Units allow a critical check on the validity of equations: all terms of an equation must have the same dimension in the kg-sec-meter SI system. Some conversion factors [2.1], [2.2] are:

| | |
|---|---|
| Length: | 1 mile = 1609.344 m, 1 yard = 0.9144 m, 1 foot = 0.3048 m, |
| Volume: | 1 UK gallon = 0.0454609 $m^3$, 1 US gallon = 0.003785 $m^3$, |
| | 1 l (liter) = $10^{-3} m^3$ = 61.03 $inch^3$ = 0.2642 US gallon, |
| Force: | 1 pound-force (lbf) = 4.44822 N, |
| Weight: | 1 pound (lb) = 0.453592 kg, |
| Pressure: | 1 pound-force per square inch = psi = $6.89474 \cdot 10^3$ Pa |
| Energy, Heat: | 1 British Thermal Unit btu = 1059.52 J (at 4°C) |

Since we have to deal mainly with water (sea water) and air we will collect the necessary properties [2.2], [2.3].

1. *Density* is usually measured in g $cm^{-3}$ or kg $m^{-3}$ or kg/l. The density $\rho$ of a medium depends on temperature and pressure and also on space and time. Sea water density depends also on the *salinity* (depending on temperature), i.e. the content of salts dissolved. Salinity ranges between 3.4 to 3.7 %. The standard value of water density is 0.999973 kg $m^{-3}$ at 4°C, 20°C: 0.99825, at 26°C 0.996785. For sea water one measures 1.02813 up to 1.03, depending on salinity. For standard air one has 1.2928 kg $m^{-3}$.

2. *Specific heat capacity* is measured in kJ $kg^{-1} K^{-1}$ or kcal $kg^{-1} K^{-1}$. The heat capacity at constant pressure $C_p$ is not equal to the capacity at constant volume $C_V$. The specific heat capacity per unit of mass $m$ is defined by $c_p = C_p/m$ and $c_V = C_V/m$. For water at 20°C one has $c_p$ 4.182 kJ $kg^{-1} K^{-1}$ or 0.999 kcal $kg^{-1} K^{-1}$. For sea water the capacity depends on salinity and ranges from 0.926 to 0.982 or 4187 J $kg^{-1} K^{-1}$. For air one has $c_V$ = 1.005 kJ $kg^{-1} K^{-1}$ = 0.240 kcal $kg^{-1} K^{-1}$ or 717 J $kg^{-1} K^{-1}$.

3. *Thermal conductivity* $\lambda$ plays an important role for the amplitude modification of *tsunamis*. For water at 20°C one has 0.598 W $m^{-1} K^{-1}$ or 0.514 kcal $m^{-1} h^{-1}$ K (h = hour). For sea water at 20°C one measures something like 0.596 W $m^{-1} K^{-1}$. For air one finds $\lambda = 0.0026$ W $m^{-1} K^{-1}$ or 0.0022 kcal $m^{-1} h^{-1} K^{-1}$. The quantity $\lambda/\rho C_p$ is called *thermal diffusivity* and is measured in $cm^2 s^{-1}$ (water: 0.0017).

4. *Viscosity* $\eta$ has an even greater influence on *tsunamis*than thermal conductivity. It depends slightly on pressure and temperature. Viscosity, also called *dynamic viscosity* (or *absolute viscosity*) , is measured in kg $m^{-1} s^{-1}$ or in *poise* (= 0.1 Pa·s or mN·s·$m^{-2}$ or dyn·s·$cm^{-2}$.) For water

at 20°C one has $\eta = 1.002$, at 30°C 0.7995 cpoise, whereas for saltwater one has about 1.075 cpoise. For air viscosity is small: $1.813 \cdot 10^{-5}$ cpoise. *Kinematic viscosity* $\nu$ is defined by $\eta/\rho$ and is measured in Stokes $\mathrm{St} = 10^{-4}\,\mathrm{m^2\,s^{-1}}$. For water at 20°C one has 0.01004 $\mathrm{cm^2\,s^{-1}}$, and for salt water at 20°C 0.01049, for air 0.143 $\mathrm{cm^2\,s^{-1}}$.

5. *Surface tension (capillarity)* $\sigma$ of water has a decisive influence on surface waves (*ripples*) on water. Water surface tension against air is 0.0727 $\mathrm{Nm^{-1}}$ ($\mathrm{kg\,s^{-2}}$) or 72.75 dyn $\mathrm{cm^{-1}}$ or $\mathrm{J\,m^{-2}}$ at 0°C.
6. *Thermal expansion* coefficients of a volume $V$ may be defined by

$$\left(\frac{1}{V}\right)\left(\frac{\partial V}{\partial T}\right)_p$$

for constant pressure $p$ (isobaric volume expansion) or by

$$-\left(\frac{V}{p}\right)\left(\frac{\partial p}{\partial V}\right)_S$$

for constant entropy $S$. These values are not of great importance in our calculations. They are very small: $20.7 \cdot 10^{-5}\,\mathrm{K^{-1}}$ for water at 20°C and $367 \cdot 10^{-5}\,\mathrm{K^{-1}}$ for air. *Compressibility* defined for constant temperature (isothermal compressibility) is given by $-(\partial V/\partial p)_T/V$.
7. *Evaporation heat* of water will be important for *hurricanes.* For water one measures 2256 $\mathrm{kJ\,kg^{-1}}$ or 538.9 kcal $\mathrm{kg^{-1}}$. Depending on the actual *saturation vapor pressure* and the temperature the *saturation humidity* (water in air) is given in Table 2.1.

Table 2.1. Saturation humidity of water in air

| water temperature | saturation kPa | pressure Torr | humidity $\mathrm{g\,m^{-3}}$ |
|---|---|---|---|
| 20°C | 2.337 | 17.53 | 17.32 |
| 25°C | 3.168 | 23.76 | 23.07 |
| 26°C | 3.361 | 25.21 | 24.40 |
| 27°C | 3.565 | 26.74 | 25.79 |

*Steam* (also *vapor*) is vaporized water, a gas interspersed with water droplets. These droplets have dimensions of 5 - 10 $\mu$ (microns) or 0.005 - 0.07 mm, in fog up to 0.1 mm. Hence steam has a white cloudy appearance. Steam is a two-phase medium. Its temperature has not to be so high as the *boiling temperature* of water. The boiling temperature depends on pressure. For water the boiling point is 100°C for normal pressure (1 at = 760 torr). But water boils at 0°C if the pressure is 4.6 Torr and it boils at 200°C at 15 at. On the other hand, evaporation takes place at temperatures lower than the boiling point. Such a *phase transition* of one one-component system is very well described by thermodynamics [2.4]. If one designates by $V_1$, the specific volume $m^3 g^{-1}$ of steam and by $V_2$ of gaseous water, then $V_1 - V_2 = \Delta V > 0$ is valid for nearly all temperatures. For 18 g water (1 mol) one has approximately $V_1 = 30{,}000$ cm$^3$ and $V_2 = 22{,}414$ cm$^3$. When the phase transition is carried out reversibly then the heat $L$ necessary for the phase transition of tepid water to the vapor phase (steam) is called *latent heat*. Then the CLAUSIUS-CLAPEYRON *equation* (vapor pressure equation) reads

$$\frac{\mathrm{d}p_{\mathrm{sat}}}{\mathrm{d}T} = \frac{L(T)}{T\Delta V}. \tag{2.1.1}$$

Here $p_{\mathrm{sat}}$ is the saturated vapor pressure (*saturation pressure)*, partial pressure of steam. Equation (2.1.1) describes the steam pressure curve $p(T)$. In order to be able to integrate (2.1.1) it is necessary to know the functions $L(T)$ and $V(T)$ or $T(V)$, respectively. Should it arrive that $L = 0$ and $\Delta V = 0$, then the concept of $p_{\mathrm{sat}}$ is meaningless. The actual values of $T$ and $p$ determine the so called critical point.

For small pressures even steam may be regarded as an ideal gas and $\Delta V$ may be replaced by $V_1 \approx V$ so that

$$p_{\mathrm{sat}} = RT/V \tag{2.1.2}$$

may be assumed. $R$ is the gas constant of the ideal gas, $R = 2$ cal K$^{-1}$ or $= 8.314510$ J mol$^{-1}$K$^{-1}$, but actually there are derivations from (2.1.2). This deviation is expressed by $Z = pV/\mathrm{RT}$ as a function of temperature and pressure. For steam at 380°K and 1 at, one finds 0.98591 [2.2]. Insertion into (2.1.1) yields

$$\frac{\mathrm{d}\ln p_{\mathrm{sat}}}{\mathrm{d}T} = \frac{L}{RT^2} \tag{2.1.3}$$

which is valid for evaporation. Here we have assumed that $L$ is a constant. Measurements of the *evaporation heat* $L$ for water (and sea water) give 539.1 cal g$^{-1}$ at 100°C and 760 Torr, but 595 cal/g ($2.49 \cdot 10^6$ J kg$^{-1}$) at 0°C. Measurements of the *evaporation rate* are difficult, they are made by the PICHE *evaporimeter.* It can be shown that the variation of $L$ with $T$ is due to the temperature dependence of the specific heat $C_p$. This allows to write down the expansion

$$L(T) = L(0°\mathrm{K}) + T \cdot (C_p\mathrm{vapor}(T) - C_{\mathrm{water}}(T))\,. \qquad (2.1.4)$$

If d$L$/d$t$ for water is known (-0.64 cal K¹ or 2680 $\mathrm{JK}^{-1}$g¹) or if (2.1.4) is generally accepted one may integrate (2.1.1). The result is

$$p_{\mathrm{sat}} = \mathrm{const} \cdot \exp\left(-\frac{L_0}{RT}\right) T^{C_{p\mathrm{vap}} - C_{p\mathrm{water}}/R}. \qquad (2.1.5)$$

The integration const depends on the substance and is sometimes called the *chemical constant.* In meteorology one uses the formula

$$p_{\mathrm{sat}} = 6.10 \cdot 10^{(7.4475t/(234.67+t))}, \qquad (2.1.6)$$

where $t =$ °C and $p_{\mathrm{sat}}$ mbar, or $p_{\mathrm{sat}} = 6.10 \cdot 10^{(8.26(T-273)/T)}$ hPa, where $T$ in Kelvin.

8. *Velocity of sound* will be a critical speed. In air of 20°C one has 344 $\mathrm{m\,s^{-1}}$, for 40°C one has 355 m $\mathrm{s}^{-1}$ and for water 1531 m $\mathrm{s}^{-1}$. The sonic speed depends on pressure and salinity. For sea water one measures 1448 up to 1620 m $\mathrm{s}^{-1}$.

## Problems

1. If one measures the saturation pressure at two temperatures, $L$ can be calculated.
   Solution: for 760 Torr, 100°C, 787.12 Torr, 101°C, $V_1 = 1674$ cm$^3$, $V_2 = 1$ cm$^3$, d$p$/d$T$ = 27.12 Torr $\mathrm{K}^{-1}$, $L$ = 538 cal $\mathrm{g}^{-1}$ or 2.26·106 J $\mathrm{kg}^{-1}$.

2. Show how the *saturation pressure* of a spherical droplet of water at 25°C depends on the radius of the droplet. ABBOT [2.4] gives the following data: at 25°C the surface tension $\sigma$ of water is 69.4 mN $\mathrm{m}^{-2}$; for a droplet hemisphere of radius $r$ the force acting as a result of the internal pressure $p_i$ is $p_i\pi r^2$ on the cut of the hemisphere. The force on this cut as a result of the external pressure $p_e$ is $p_e\pi r^2$. Hence one has the force balance

$$p_i\pi r^2 - p_e\pi r^2 = \sigma 2\pi r \quad \text{or} \quad p_i - p_e = 2\sigma/r. \qquad (2.1.7)$$

   For a pressure difference $\Delta p$ in Pa and $r$ in m the solution is

| $\Delta p$: | 14 | 1390 | 138800 |
|---|---|---|---|
| $r$: | 0.01 | 0.0001 | 0.000001 |

## 2.2 Conservation of mass

When dealing with tsunamis and hurricanes we actually are concerned with three substances: air, water and vapor. The mass of air may stay outside considerations, since it remains constant. Contrary to this we must take into account a certain phase transition (evaporation, condensation) between the two phases water and vapor. Large amounts of sea water will be available, but the amount of vapor in hurricanes will change according to temperature and other factors.

Let $\rho(x, y, z, t)$ be the actual density of vapor, then the mass $M$ of vapor contained in the volume $V$ will be given by

$$M = \int_V \rho(x, y, z, t) \cdot dxdydz. \tag{2.2.1}$$

Two phenomena may vary $M$:

1. Outflow (inflow) through the surface $F$ enclosing the volume $V$ and
2. additional production $D$ of vapor within the domain $V$.

If one defines the outflow by $\int_F \rho\vec{u}d\vec{f}$, then the mass conservation is defined by [2.5]

$$-\frac{\partial}{\partial t}\int_V \rho(x, y, z, t)dxdydz = \int_F \rho\vec{u}d\vec{f} + \int_V D(x, y, z, t)dxdydz. \tag{2.2.2}$$

Using the GAUSS *theorem*

$$\int_F \rho\vec{u}d\vec{f} = \int_V \text{div}(\rho\vec{u})dxdydz \tag{2.2.3}$$

one obtains the *continuity equation*

$$\frac{\partial\rho}{\partial t} + \text{div}(\rho\vec{u}) = D. \tag{2.2.4}$$

$D$ is to be measured in kg $\text{m}^{-3}\text{s}^{-1}$ and may be called *evaporation rate*, $\vec{u}$ is the stream velocity and div$\vec{A}$ of a vector $\vec{A}(x, y, z, t)$ is defined by [2.6]

$$\frac{\partial A_x}{\partial x} + \frac{\partial A_y}{\partial y} + \frac{\partial A_z}{\partial z} = \nabla\vec{A}. \tag{2.2.5}$$

For our purposes the continuity equation is of interest in the one-dimensional nonsteady case in Cartesian (and later in circular cylindrical) coordinates

$$\frac{\partial \rho}{\partial t} + \rho \frac{\partial u}{\partial x} + u \frac{\partial \rho}{\partial x} = D(x,t). \tag{2.2.6}$$

The source term $D$ depends on pressure $p(x,t)$ and temperature $T(x,t)$, but it may be written as $D(x,t)$. It can be expressed by the evaporation rate depending on salinity, wind etc.

Whereas the source term is important for hurricanes it plays no role in the investigation of tsunamis. These large nonlinear waves exhibit *amplitude modulation.* Some calculations have indicated that new dependent variables are of advantage. Hence the specific volume $s$ has been used instead of the density $\rho$. Insertion of

$$s(x,t) = 1/\rho(x,t) \tag{2.2.7}$$

into the sourcefree continuity equation yields

$$-s_t + su_x - us_x = 0. \tag{2.2.8}$$

This is a quasilinear partial differential equation of first order.

In a one-dimensional nonsteady flow problem of a compressible viscous substance one has to determine the stream velocity $u(x,t)$, the density $\rho(x,t)$ or the specific volume $s(x,t)$, the pressure $p(x,t)$ and the temperature $T(x,t)$. Hence one needs 5 partial differential equations of first order or one partial differential equation of fifth order. The equation of motion, the energy theorem and two thermodynamic relations, the general equation of state $F(p,\rho,T)$ and a change of state are needed additionally. In principle, such a state of change could be isobaric, isochoric, isentropic, isenthalpic, isothermal and reversible adiabatic. In some cases (viscous substances) irreversible thermodynamics has to be taken into account.

For a sourcefree incompressible fluid ($\rho = \text{const}$) the continuity equation becomes $\operatorname{div} \vec{u} = 0$, so that the vector $\vec{u}$ may be represented by the curl of another vector $\vec{A}$, $\vec{u} = \operatorname{curl} \vec{A}$, since by definition $\operatorname{div} \operatorname{curl} \vec{A} = 0$ for any vector. If $\operatorname{curl} \vec{A} = 0$, then $\vec{A}$ may be calculated from a scalar $\vec{A} = \nabla U$.

## Problems

1. Transform (2.2.4) into cylindrical coordinates [2.6].

2. Calculate $D$ for a cylinder of height 1000 m, if 7 mm water will be evaporated per $m^2$ during 24 hours.

Solution: According to (2.2.6) $D$ has the dimension of $\partial\rho/\partial t$. We find from table 2.2 that air at 26° C contains 24.4 g m$^{-3}$ water-vapor. How long would it take that the evaporation rate $D = 0.814{\cdot}10^{-10}$ g cm$^{-3}$ s$^{-1}$ produces this humidity? ($30 \cdot 10^4$ s).

3. Derive (2.2.8) from (2.2.7) and (2.2.6) for $D = 0, D \neq 0$.

## *2.3 The equation of motion*

The linear momentum density of a liquid is given by $\rho v$ kg m$^{-2}$ s$^{-1}$. In a closed system (no exterior forces) the momentum is conserved. Neglecting at first all exterior influences, we consider possible changes $\delta\vec{v}$ of the vector field $\vec{v}$ due to surface forces. We expand into four terms:

$$\vec{v} = \vec{v}_0 + \delta_1\vec{v} + \delta_2\vec{v} + \delta_3\vec{v}. \tag{2.3.1}$$

The designations of these four terms will be:
$\vec{v}_0$ = pure translation of small fluid volumes due to internal pressure,
$\delta_1\vec{v}$ = rigid rotation of a fluid volume,
$\delta_2\vec{v}$ = expansion or compression of a fluid element without modification of its shape due to internal compressional forces,
$\delta_3\vec{v}$ = deformation of the shape at constant volume (non-dilatational strain) due to internal shear forces.
Collecting these terms, we may write

$$\vec{v} = \vec{v}_0 + \nabla;\vec{v}. \tag{2.3.2}$$

Here $\nabla; v$ is the tensor product of the two vectors $\nabla$ and $\vec{v}$. The tensor in (2.3.2) is called *strain tensor.* In Cartesian coordinates the tensor product may be written

$$\nabla;\vec{v} = \begin{pmatrix} \dfrac{\partial v_x}{\partial x} & \dfrac{\partial v_y}{\partial x} & \dfrac{\partial v_z}{\partial x} \\[2ex] \dfrac{\partial v_x}{\partial y} & \dfrac{\partial v_y}{\partial y} & \dfrac{\partial v_z}{\partial y} \\[2ex] \dfrac{\partial v_x}{\partial z} & \dfrac{\partial v_y}{\partial z} & \dfrac{\partial v_z}{\partial z} \end{pmatrix}. \tag{2.3.3}$$

In order to identify the changes defined in (2.3.1) we first define the reverted tensor

$$\nabla\tilde{;}\vec{v} = \begin{pmatrix} \frac{\partial v_x}{\partial x} & \frac{\partial v_x}{\partial y} & \frac{\partial v_x}{\partial z} \\ \frac{\partial v_y}{\partial x} & \frac{\partial v_y}{\partial y} & \frac{\partial v_y}{\partial z} \\ \frac{\partial v_z}{\partial x} & \frac{\partial v_z}{\partial y} & \frac{\partial v_z}{\partial z} \end{pmatrix}. \tag{2.3.4}$$

Then we split (2.3.2) up

$$\begin{aligned} \vec{v} = \quad & \vec{v}_0 + \nabla; v = \vec{v}_0 + \frac{1}{2}(\nabla;\vec{v} - \nabla\tilde{;}\vec{v}) \\ & + \frac{1}{2}(\nabla;\vec{v} + \nabla\tilde{;}\vec{v}) - \frac{1}{3}\mathrm{div}\,\vec{v}\cdot\vec{E} \\ & + \frac{1}{3}\mathrm{div}\,\vec{v}\cdot\vec{E}. \end{aligned} \tag{2.3.5}$$

Here $\vec{E}$ is the unit tensor

$$\vec{E} = \begin{pmatrix} 1 & 0 & 0 \\ 0 & 1 & 0 \\ 0 & 0 & 1 \end{pmatrix} = \delta_{ik}. \tag{2.3.6}$$

It is easy to see that the second right-hand term in the first line of (2.3.5) corresponds to a rigid rotation

$$\delta_1\vec{v} = \frac{1}{2}\begin{pmatrix} 0 & \frac{\partial v_y}{\partial x} - \frac{\partial v_x}{\partial y} & \frac{\partial v_z}{\partial x} - \frac{\partial v_x}{\partial z} \\ \frac{\partial v_x}{\partial y} - \frac{\partial v_y}{\partial x} & 0 & \frac{\partial v_z}{\partial y} - \frac{\partial v_y}{\partial z} \\ \frac{\partial v_x}{\partial z} - \frac{\partial v_z}{\partial x} & \frac{\partial v_y}{\partial v_z} - \frac{\partial z}{\partial y} & 0 \end{pmatrix} \tag{2.3.7}$$

which may be written as

$$\delta_1\vec{v} = \begin{pmatrix} 0 & +\frac{1}{2}(\mathrm{curl}\,\vec{v})_z & -\frac{1}{2}(\mathrm{curl}\,\vec{v})_y \\ -\frac{1}{2}(\mathrm{curl}\,\vec{v})_z & 0 & \frac{1}{2}(\mathrm{curl}\,\vec{v})_x \\ \frac{1}{2}(\mathrm{curl}\,\vec{v})_y & -\frac{1}{2}(\mathrm{curl}\,\vec{v})_x & 0 \end{pmatrix}. \tag{2.3.8}$$

The second line in (2.3.5) corresponds to a deformation of the shape

$$\delta_3\vec{v}=\begin{pmatrix} \frac{2}{3}\frac{\partial v_x}{\partial x}-\frac{1}{3}\frac{\partial v_y}{\partial y}-\frac{1}{3}\frac{\partial v_z}{\partial z} & \frac{1}{2}\left(\frac{\partial v_y}{\partial x}+\frac{\partial v_x}{\partial y}\right) & \frac{1}{2}\left(\frac{\partial v_z}{\partial x}+\frac{\partial v_x}{\partial z}\right) \\ \frac{1}{2}\left(\frac{\partial v_y}{\partial x}+\frac{\partial v_x}{\partial y}\right) & \frac{2}{3}\frac{\partial v_y}{\partial y}-\frac{1}{3}\frac{\partial v_x}{\partial x}-\frac{1}{3}\frac{\partial v_z}{\partial z} & \frac{1}{2}\left(\frac{\partial v_z}{\partial y}+\frac{\partial v_y}{\partial z}\right) \\ \frac{1}{2}\left(\frac{\partial v_z}{\partial x}+\frac{\partial v_x}{\partial z}\right) & \frac{1}{2}\left(\frac{\partial v_y}{\partial z}+\frac{\partial v_z}{\partial y}\right) & \frac{2}{3}\frac{\partial v_z}{\partial z}-\frac{1}{3}\frac{\partial v_x}{\partial x}-\frac{1}{3}\frac{\partial v_y}{\partial y} \end{pmatrix}. \tag{2.3.9}$$

The tensor (2.3.9) is called *deformation tensor.* The last line in (2.3.5) describes $\delta_2\vec{v}$.

Now we consider the forces (stresses) acting upon the surface of a fluid element. In first approximation strain and stress are proportional one to each other. Due to the conservation of angular momentum the stress tensor $\Pi$ is symmetric, $\Pi_{ik} = \Pi_{ki}$. A further consequence is that no term of the *stress tensor* appears which is proportional to $\delta_1\vec{v}$. We thus make the setup [2.2], [2.6] (which can be proven by statistical physics [2.8])

$$\Pi_{ik} = -p\cdot\delta_{ik} + 2\eta\cdot\frac{1}{2}\left(\frac{\partial v_i}{\partial x_k}+\frac{\partial v_k}{\partial x_i}\right) - \frac{1}{3}\eta'\operatorname{div}\vec{v}\cdot\delta_{ik} \tag{2.3.10}$$

corresponding to: $\vec{v}_0$ $\delta_3\vec{v}$ $\delta_2\vec{v}$.
This setup contains two parameters $\eta$ and $\eta'$. Here $\eta$ is again the *absolute* (dynamic) *viscosity* and $\eta'$ is the *dilatational viscosity* (*compressional viscosity*) .

As for the physical units we have

$$\Pi \to p \to \mathrm{kg\,m^{-1}\,s^{-2}} \to \eta\frac{\partial v}{\partial x} \to \mathrm{kg\,m^{-1}\,s^{-1}\cdot m\,s^{-1}\,m^{-1}}.$$

Thus $\eta'$ is expressed in the same units as $\eta$.

The compressional viscosity $\eta'$ is very small for many substances and it is thence very often neglected. In the past, the STOKES *hypothesis* for the vanishing *bulk viscosity*

$$\eta' + \frac{2}{3}\eta = 0 \quad (\text{also } \eta' - \frac{2}{3}\eta = 0) \tag{2.3.11}$$

has been assumed. For monoatomic gases one has $2\eta = -3\eta'$. In statistical physics some arguments can be given for this assumption. For incompressible substances one has $\operatorname{div}\vec{v} = 0$ and no STOKES hypothesis is necessary to arrive at $\operatorname{div}\vec{v} = 0$. In order to understand the consequence of (2.3.11) on (2.3.10) we rewrite this equation

$$\Pi_{ik} = -p\delta_{ik} + \eta\left(\frac{\partial v_i}{\partial x_k} + \frac{\partial v_k}{\partial x_i}\right). \tag{2.3.12}$$

If $i \neq k$ the *stress tensor* $\Pi_{ik}$ describes shearing stress and the diagonal terms $\Pi_{ii}$ represent compresssion.

We now may discuss the conservation of momentum and derive the equation of motion of a fluid volume element $\mathrm{d}\tau$. According to NEWTON's law one has

$$\begin{aligned} \rho \mathrm{d}\tau \frac{\mathrm{d}\vec{v}}{\mathrm{d}t} &= -\nabla p \mathrm{d}\tau + \rho \mathrm{d}\tau \vec{g} + \eta\Delta\vec{v}\cdot \mathrm{d}\tau, \\ \frac{\partial\vec{v}}{\partial t} + (\vec{v}\,\nabla)\vec{v} &= -\frac{\nabla p}{\rho} + \vec{g} + \nu\Delta\vec{v}, \end{aligned} \tag{2.3.13}$$

where $\nu$ is the *kinematic viscosity* $\eta/\rho$ which depends on $\vec{x}$ and $t$ due to $\rho(\vec{x}, t)$. We assume that $\eta$ and $\nu$ are constant. (For water one has $\eta = 0.010\,\mathrm{g\,cm^{-1}\,s^{-1}}$, $\nu = 0.010\,\mathrm{cm^2\,s^{-1}}$ and for air $\eta = 1.8 \cdot 10^{-4}$, $\nu = 0.150$).

Equation (2.3.13) is called the NAVIER-STOKES *equation.* If bulk viscosity is included, it reads

$$\rho\left(\frac{\partial\vec{v}}{\partial t} + \nabla\frac{v^2}{2} - \vec{v}\times(\nabla\times\vec{v})\right) = -\nabla p + \rho\vec{g} - \nabla\times[\eta(\nabla\times\vec{v})] + \nabla((\eta' + 2\eta)\nabla\cdot\vec{v}). \tag{2.3.14}$$

Here the identity $(\vec{v}\,\nabla)\vec{v} = \nabla(v^2/2) - \vec{v}\times(\nabla\times\vec{v})$ has been used.

Exterior forces are gravity $\vec{g}$ and the CORIOLIS *force* . Assuming that a potential $V$ can represent the gravity forces we define

$$\vec{g} = -\nabla V. \tag{2.3.15}$$

If viscosity is neglected then (2.3.13) assumes the form

$$\frac{\partial\vec{v}}{\partial t} + (\vec{v}\,\nabla)\vec{v} = -\nabla P - \nabla V, \tag{2.3.16}$$

where

$$P = \int\frac{\mathrm{d}p}{\rho(p)}, \quad \frac{\nabla p}{\rho} = \nabla P, \tag{2.3.17}$$

where $P$ is called *pressure density integral.* It can be calculated, if $\rho(p)$ is given, for instance for an ideal isothermal gas $\rho = \mathrm{const}\, p$. In two cases

1. $\operatorname{curl}\vec{v} = 0$ (potential flow),

2. $\vec{v}||\text{curl}\,\vec{v}$, $\vec{v} \times \text{curl}\,\vec{v} = 0$ (BELTRAMI *flow*)

equation (2.3.14) may be written in the form

$$\frac{\partial \vec{v}}{\partial t} = -\nabla \left( \frac{\vec{v}^{\,2}}{2} + P + V \right). \tag{2.3.18}$$

For water one may assume $\rho$ = const. Then one obtains

$$\rho \frac{\partial \vec{v}}{\partial t} = -\nabla \left( \frac{\rho \vec{v}^{\,2}}{2} + p + V\rho \right). \tag{2.3.19}$$

For a steady flow $\partial/\partial t = 0$ and $\rho V = \rho g h$ (*hydrostatic pressure* due to gravity) one obtains after integration

$$\frac{\rho \vec{v}^{\,2}}{2} + p + \rho g z = \text{const} = \frac{\rho \vec{v}_1^{\,2}}{2} + p_1 + \rho g z_1. \tag{2.3.20}$$

Here $z$ is the water depth and $p$ is the pressure. The CORIOLIS force has not been included. Equation (2.3.20) is called BERNOULLI *equation.* Another form of the BERNOULLI equation will be discussed in section 2.7.

Now the speed of efflux from an opening in a reservoir equals the speed $v$ of the liquid would acquire if falling from rest $v_1 = 0$ from the surface $p_1 = p_0$ of the reservoir down to the opening $p = p_0$ (TORICELLI's *theorem*). Using $z_1 - z = h$, the theorem reads

$$v = \sqrt{2gh}. \tag{2.3.21}$$

This formula will be of value when we discuss the equivalence theorem in section 2.10.

## Problems

1. Calculate the pressure density integral (2.3.17) for isothermal conditions $p_1/\rho_1 = p_2/\rho_2$ = const = $RT$ and insert it into (2.3.21) for steady flow ($\partial/\partial t = 0$).

   Solution [2.7]:

$$\frac{p_1}{\rho_1} \ln p_1 + \frac{v_1^2}{2g} + z_1 = \frac{p_2}{\rho_2} \ln p_2 + \frac{v_2^2}{2g} + z_2. \tag{2.3.22}$$

2. Calculate the shape of the surface of water which is contained in a rotating cylindrical container.

   Solution: $v_x = -\omega y$, $v_y = \omega x$, $v_z = 0$. The equation of motion yields

$$x\omega^2 = \frac{1}{\rho}\frac{\partial p}{\partial x}, \quad y\omega^2 = \frac{1}{\rho}\frac{\partial p}{\partial y}, \quad \frac{1}{\rho}\frac{\partial p}{\partial z} + g = 0. \tag{2.3.23}$$

Integration yields

$$\frac{p}{\rho} = \frac{\omega^2}{2}(x^2 + y^2) - gz + \text{constant}. \tag{2.3.24}$$

Since the free surface of water is defined by $p = \text{const}$, the shape of the surface is described by the paraboloid

$$z = \frac{\omega^2}{2g}(x^2 + y^2). \tag{2.3.25}$$

3. Derive the NAVIER-STOKES equation in circular cylinder coordinates $r$, $\vartheta$, $z$. Assume that the viscosity $= \eta(r, z, \vartheta)$.

   Solution [2.6] ($\vec{F}$ are the exterior body forces):

$$\frac{D}{Dt} = \frac{\partial}{\partial t} + v_r \frac{\partial}{\partial r} + \frac{v_\vartheta}{r}\frac{\partial}{\partial \vartheta} + v_z \frac{\partial}{\partial z}, \tag{2.3.26}$$

$$\nabla \cdot \vec{v} = \frac{1}{r}\frac{\partial}{\partial r}(r v_r) + \frac{1}{r}\frac{\partial v_\vartheta}{\partial \vartheta} + \frac{\partial v_z}{\partial z} = \frac{\partial v_r}{\partial r} + \frac{v_r}{r} + \frac{1}{r}\frac{\partial v_\vartheta}{\partial \vartheta} + \frac{\partial v_z}{\partial z} = \operatorname{div} \vec{v}. \tag{2.3.27}$$

$$\begin{aligned} \rho\left[\frac{Dv_r}{Dt} - \frac{v_\vartheta^2}{r}\right] &= F_r - \frac{\partial p}{\partial r} + \frac{\partial}{\partial r}\left[2\eta\frac{\partial v_r}{\partial r} + \left(\eta' - \frac{2}{3}\eta\right)\nabla \cdot \vec{v}\right] \\ &+ \frac{1}{r}\frac{\partial}{\partial \vartheta}\left[\eta\left(\frac{1}{r}\frac{\partial v_r}{\partial \vartheta} + \frac{\partial v_\vartheta}{\partial r} - \frac{v_\vartheta}{r}\right)\right] + \frac{\partial}{\partial z}\left[\eta\left(\frac{\partial v_r}{\partial z} + \frac{\partial v_z}{\partial r}\right)\right] \\ &+ \frac{2\eta}{r}\left(\frac{\partial v_r}{\partial r} - \frac{1}{r}\frac{\partial v_\vartheta}{\partial \vartheta} - \frac{v_r}{r}\right), \end{aligned} \tag{2.3.28}$$

$$\begin{aligned} \rho\left[\frac{Dv_\vartheta}{Dt} + \frac{v_r v_\vartheta}{r}\right] &= F_\vartheta - \frac{1}{r}\frac{\partial p}{\partial \vartheta} + \frac{1}{r}\frac{\partial}{\partial \vartheta}\left[\frac{2\eta}{r}\frac{\partial v_\vartheta}{\partial \vartheta} + \left(\eta' - \frac{2}{3}\eta\right)\nabla \cdot \vec{v}\right] \\ &+ \frac{\partial}{\partial z}\left[\eta\left(\frac{1}{r}\frac{\partial v_z}{\partial \vartheta} + \frac{\partial v_\vartheta}{\partial z}\right)\right] + \frac{\partial}{\partial r}\left[\eta\left(\frac{1}{r}\frac{\partial v_r}{\partial \vartheta} + \frac{\partial v_\vartheta}{\partial r} - \frac{v_\vartheta}{r}\right)\right] \\ &+ \frac{2\eta}{r}\left[\frac{1}{r}\frac{\partial v_r}{\partial \vartheta} + \frac{\partial v_\vartheta}{\partial r} - \frac{v_\vartheta}{r}\right], \end{aligned} \tag{2.3.29}$$

$$\begin{aligned} \rho\frac{Dv_z}{Dt} &= F_z - \frac{\partial p}{\partial z} + \frac{\partial}{\partial z}\left[2\eta\frac{\partial v_z}{\partial z} + \left(\eta' - \frac{2}{3}\eta\right)\nabla \cdot \vec{v}\right] \\ &+ \frac{1}{r}\frac{\partial}{\partial r}\left[\eta r\left(\frac{\partial v_r}{\partial z} + \frac{\partial v_z}{\partial r}\right)\right] + \frac{1}{r}\frac{\partial}{\partial \vartheta}\left[\eta\left(\frac{1}{r}\frac{\partial v_z}{\partial \vartheta} + \frac{\partial v_\vartheta}{\partial z}\right)\right]. \end{aligned} \tag{2.3.30}$$

4. If there are sources of gas $D$, see (2.2.6), then it may be necessary to modify the equation of motion. Evaporation of streaming droplets

of water in a gas flow with velocity $v$ will give rise to an equation of motion [2.5] in the form

$$\rho\frac{\partial v}{\partial t}+\rho v\frac{\partial v}{\partial x}+\frac{\partial p}{\partial x}=-D(x,t)v. \tag{2.3.31}$$

Assume that $D$ does not depend on pressure and temperature and derive a BERNOULLI *equation* from (2.3.31) for a steady potential flow of a compressible medium ($v=-\partial\varphi/\partial x, V=0$).

Solution: $v^2/2+P=\int(D/\rho)\mathrm{d}x$.

## *2.4 Conservation of energy*

In point mechanics it is easy to derive the energy conservation from the equation of motion. Multiplication of the equation of motion by $\vec{v}$ and integration yields the conservation of the sum of kinetic and potential energy of a mass point. In fluid mechanics the situation is more complicated since viscosity, compression, heat conduction etc play a certain role. Actually the BERNOULLI *equation* (2.3.20) is an energy principle, but restricted to an incompressible steady nonviscous flow. If the medium is compressible but still nonviscous, then the BERNOULLI *equation* (2.3.22) for a pure gravity field in the $z$-direction reads

$$\int\frac{\mathrm{d}p}{\rho(p)}+\frac{v^2}{2}+gz=\text{const.} \tag{2.4.1}$$

Due to $\rho=\rho(p)$ either thermodynamics enters the system of equations or isothermal processes must be assumed. If viscosity, heat conduction and heat loss have to be taken into account, then the energy balance will be quite complicated. We choose the following designations: internal energy per unit mass $U$ (which contains $\rho v^2/2$), the internal heat loss $Q$, for instance due to evaporation, thermal conductivity $\lambda$ which may be variable, $\Phi$(x,y,z,t) is the dissipation function expressing the energy dissipation rate per unit volume (or unit mass) at point $(x,y,z,t)$ and $T$ is the temperature. The setup for the thermodynamic *energy balance, first law of thermodynamics* (which can be derived from statistical physics) is given [2.6] by

$$\rho\left(\frac{\partial U}{\partial t}+\vec{v}\cdot\nabla U\right)+p\nabla\cdot\vec{v}=\frac{\partial Q}{\partial t}+\Phi+\nabla(\lambda\nabla T). \tag{2.4.2}$$

Exterior forces and mechanical terms have not been taken into account. The internal energy $U$ will be discussed in the next section. With the help of the continuity equation (2.2.4) which we use in the form

$$\frac{\partial \rho}{\partial t} + \rho \nabla \cdot \vec{v} + \vec{v} \nabla \rho = D, \tag{2.4.3}$$

the third left-hand side term of (2.4.2) may be written as

$$p \nabla \cdot v = p \left( \frac{D}{\rho} - \vec{v} \frac{\nabla \rho}{\rho} - \frac{1}{\rho} \frac{\partial \rho}{\partial t} \right). \tag{2.4.4}$$

Now a new variable namely $T$ appears in the equations. The function $D(x,t)$ must be known or has to be calculated $D(p,T)$ from thermodynamic considerations. It may be connected with $Q$ which also depends on thermodynamic considerations. The dissipation function $\Phi$ is determined by the setup of the *stress tensor* (2.3.10). In the literature not only (2.3.10) but also other setups are used. Thus the expression for $\Phi$ may vary a little. Without use of the STOKES *hypothesis* (2.3.11) one obtains [2.6]

$$\Phi = 2\eta \left[ \left( \frac{\partial u}{\partial x} \right)^2 + \left( \frac{\partial v}{\partial y} \right)^2 + \left( \frac{\partial w}{\partial z} \right)^2 + \frac{1}{2} \left( \frac{\partial u}{\partial y} + \frac{\partial v}{\partial x} \right)^2 + \frac{1}{2} \left( \frac{\partial v}{\partial z} + \frac{\partial w}{\partial y} \right)^2 \right.$$
$$\left. + \frac{1}{2} \left( \frac{\partial w}{\partial x} + \frac{\partial u}{\partial z} \right)^2 \right] + \eta' \left( \frac{\partial u}{\partial x} + \frac{\partial v}{\partial y} + \frac{\partial w}{\partial z} \right)^2 \tag{2.4.5}$$

or [ ] using the STOKES hypothesis

$$\Phi = 2\eta \left[ u_x^2 + v_y^2 + w_z^2 \right] - \frac{2}{3} \eta \left( u_x + v_y + w_z \right)^2$$
$$+ \eta \left( u_y + v_x \right)^2 + \eta \left( u_z + w_x \right)^2 + \eta \left( v_z + w_y \right)^2, \tag{2.4.6}$$

where we used the designation $u$, $v$, $w$ for the components of $\vec{v}$. Another setup of the stress tensor yields [2.9]

$$\Phi = \eta \left[ 2 \left( u_x^2 + v_y^2 + w_z^2 \right) + \left( u_y + v_x \right)^2 + \left( u_z + w_x \right)^2 + \left( v_z + w_y \right)^2 \right], \tag{2.4.7}$$

whereas other authors prefer for an incompressible fluid the simple expression [2.10]

$$\Phi = \frac{\eta}{2} \left( \frac{\partial v_i}{\partial x_k} + \frac{\partial v_k}{\partial x_i} \right)^2. \tag{2.4.8}$$

The dissipation reduces the kinetic energy of the fluid.

For the evaporation $E$ from free water surfaces measurements give 1000 up to 3000 mm per year, depending on the water temperature. The most simple setup is given by an empirical formula by DALTON

$$E = (0.13 + 0.094v)(p_{\mathrm{Sat}} - p_{\mathrm{act}}), \tag{2.4.9}$$

where $E$ in mm per year, $v$ is the velocity of wind ($\mathrm{m\,s^{-1}}$) over the water surface. The theoretical value of $p_{\mathrm{Sat}}$ may be calculated from (2.1.5) or (2.1.6). $p_{\mathrm{act}}$ is the actual vapor pressure. The pressures are given in hPa ($10^2$ Pa). The heat consumption rate $\partial Q/\partial t$ may be calculated from the evaporation rate, see problem 3.

A look on the basic equations shows the following situation: in sections 2.2 and 2.3 we were concerned with three basic equations: the continuity equation (2.2.4), the equation of motion for a non-viscous, non-conducting but compressible isothermal fluid satisfying (2.3.17) and the BERNOULLI equation or the equation of motion for a nonviscous non-conducting ideal gas. We had to do with the equations of continuity, of motion and the energy theorem, which described three physical quantities: velocity $\vec{v}$, pressure $p$ and density $\rho$.

As soon as a fluid has to be regarded as compressible to be not isothermal, or if dissipation (due to viscosity, heat conduction etc) has to be taken into account, thermodynamics comes into the play. Is a non-dissipative flow isothermal? Is it possible to use the state equation for an ideal gas? How does viscosity and/or heat conduction change the situation? We will discuss these questions in the next section.

## Problems

1. Transform the energy equation (2.4.2) into cylindrical coordinates $r, \vartheta, z$. Solution:

$$\begin{gathered}\rho\left(\frac{\partial U}{\partial t} + \vec{v}\nabla U\right) + p\,\mathrm{div}\,\vec{v} = \\ \frac{\partial Q}{\partial t} + \Phi + \frac{1}{r}\frac{\partial}{\partial r}\left(r\lambda\frac{\partial T}{\partial r}\right) + \frac{1}{r^2}\frac{\partial}{\partial \vartheta}\left(\lambda\frac{\partial T}{\partial \vartheta}\right) + \frac{\partial}{\partial z}\left(\lambda\frac{\partial T}{\partial z}\right).\end{gathered} \tag{2.4.10}$$

   Specify $\vec{v}\nabla U$, $p\,\mathrm{div}\,\vec{v}$ and write (2.4.10) for constant $\lambda$ and an incompressible fluid.

2. Transform the dissipation function (2.4.5) into cylindrical coordinates.

Solution:

$$\Phi = \eta\left[2\left\{\left(\frac{\partial v_r}{\partial r}\right)^2 + \left(\frac{1}{r}\frac{\partial v_\vartheta}{\partial \vartheta} + \frac{\partial v_r}{r}\right)^2 + \left(\frac{\partial v_z}{\partial z}\right)^2\right\}\right.$$
$$+\left(\frac{1}{r}\frac{\partial v_z}{\partial \vartheta} + \frac{\partial v_z}{\partial z}\right)^2 + \left(\frac{\partial v_r}{\partial z} + \frac{\partial v_z}{\partial r}\right)^2 + \left.\left(\frac{1}{r}\frac{\partial v_r}{\partial \vartheta} + \frac{\partial v_\vartheta}{\partial r} - \frac{v_\vartheta}{r}\right)^2\right]$$
$$+\eta'\left[\frac{\partial v_r}{\partial r} + \frac{1}{r}\frac{\partial v_\vartheta}{\partial \vartheta} + \frac{v_r}{r} + \frac{\partial v_z}{\partial z}\right]^2. \tag{2.4.11}$$

3. What evaporation energy rate $(\partial Q/\partial t)$ is necessary to produce an evaporation density rate $D = 0.814 \cdot 10^{-10}\,\mathrm{g\,cm}^{-3}\,\mathrm{s}^{-1}$, see problem 2 in section 2.2. The solution is simple, if the dependence of $L(T)$ on $T$ is neglected. In this case one may assume that $2.49{\cdot}10^6\,\mathrm{J\,kg}^{-1}$ are necessary to evaporate 1 kg water. If $L(T)$ has to be taken into account, then (2.1.4) has to be used.

## *2.5 Thermodynamics*

The three basic flow equations discussed in the last sections use four variables: velocity $\vec{v}$, density $\rho$, pressure $p$, temperature $T$ and a still unknown relation $\rho(p)$ and the explicit expression for the internal energy $U$. We thus need two more equations.

First of all we discuss *equations of state*

$$p = f(\rho, V). \tag{2.5.1}$$

For an ideal gas (defined by statistical physics) the BOYLE-MARIOTT-GAY-LUSSAC law is valid

$$pV = nRT. \tag{2.5.2}$$

Here $V$ is the volume of a mass $M$ of gas consisting of $n$ mol. In units we have:

$$\begin{array}{ccccccc} p & \cdot\ V & = n & \cdot & R & \cdot & T \\ \mathrm{at} = \mathrm{kg\,s^{-2}\,m^{-1}} & \cdot\ \mathrm{m^3} & \mathrm{mol} & \cdot & \mathrm{kg\,s^{-2}\,m^2\cdot mol^{-1}\,K^{-1}} & \cdot & \mathrm{K}, \end{array} \tag{2.5.3}$$

where $\mathrm{kg\,s^{-2}\,m^2 = J = N\ m}$. $R = 8.3144\,\mathrm{J\,mol^{-1}\,K^{-1}}$ is the universal gas constant.

Due to $V\rho = M$ one has for one mol ($n = 1$) $M = \mu$, where $\mu$ is the molecular weight, for water 18 g, V = 22.414 l. Then (2.5.2) becomes

$$p = \rho CT, \tag{2.5.4}$$

where the *specific gas constant* $C$ in $\mathrm{m^2\,s^{-2}\,K^{-1}}$ is given by

$$C = \frac{R}{M} = \frac{R}{\mu}. \tag{2.5.5}$$

For dry air one has $287\,\mathrm{J\,kg^{-1}\,K^{-1}}$.

The fluid in hurricanes is steam which is however not an ideal gas. For low pressures steam may be described by the empirical CALLENDAR *equation*

$$pV = RT + \left(b - a^2/T^n\right) p, \tag{2.5.6}$$

where $a, b, n$ are constants specific for the steam.

A more elaborate model for a real gas is given by the VAN DER WAALS *equation*

$$\left(p + a/V^2\right)(V - b) = RT. \tag{2.5.7}$$

The new parameters $a$ (*cohesion pressure*) and $b$ (*covolume*, excluded volume) are constants depending on the particular gas. The parameter $a$ describes an intermolecular attractive force and $b$ accounts for the finite moleculare size. The isotherms $T =$ const are polynomials of third degree. In order to be independent from the parameters $a$ and $b$ we first determine the *critical point* of (2.5.7). This point is defined by

$$\left(\frac{\partial p}{\partial V}\right)_T = 0, \quad \left(\frac{\partial^2 P}{\partial V^2}\right)_T = 0. \tag{2.5.8}$$

Solving (2.5.7) and (2.5.8) for $p, V$ and $T$, the critical point gives

$$p_c = a/27b^2, \quad V_c = 3b \;[= 3bm \text{ for (2.5.14)}], \quad T_c = 8a/27\,Rb. \tag{2.5.9}$$

If the critical values have been measured, one may calculate the gas specific values $a$ and $b$. Equation (2.5.9) shows that the constants $a$ and $b$ are overdetermined since there are three equations and only two unknowns. This implies that all three critical properties cannot be satisfied by (2.5.7). It is advantageous to express $a$ and $b$ in terms of two critical values. Elimination of $V_c$ from (2.5.7) and (2.5.9) yields

$$b = RT_c/8p_c, \quad a = 27R^2T_c^2/64p_c. \tag{2.5.10}$$

Now the VAN DER WAALS equation (2.5.7) can be written in terms of reduced variables defined by

$$\bar{p} = p/p_c, \tag{2.5.11}$$

$$\bar{V} = V/V_c, \quad \bar{T} = T/T_c. \tag{2.5.12}$$

We thus obtain

$$\bar{p} = 8\bar{T}/(3\bar{V} - 1) - 3/\bar{V}^2. \tag{2.5.13}$$

Now the reduced VAN DER WAALS equation (2.5.13) may be plotted and discussed without the knowledge of the specific values $a$ and $b$.

For our problem to investigate the equation of state for steam, we rewrite the VAN DER WAALS *equation* (2.5.7) in a form considering a steam mass $m$ kg

$$\left(p + \frac{am^2}{V^2}\right)(V - bm) = mCT. \tag{2.5.14}$$

The pressure $p$ is again measured in $\mathrm{N\,m^{-2}}$ ($= \mathrm{kg\,s^{-2}\,m^{-1}}$), the gas constant $C$ is given in $\mathrm{J\,kg^{-1}\,K^{-1}}$ ($= \mathrm{m^2\,s^{-2}\,K^{-1}}$), $a$ is measured in $\mathrm{N\,m^4\,kg^{-2}}$ ($= \mathrm{m^5\,kg^{-1}\,s^{-2}}$) and $b$ in $\mathrm{m^3\,kg^{-1}}$, since $V$ is measured in $\mathrm{m^3}$. One sees that each term in (2.5.14) is measured in Joule. The critical values (2.5.9) for steam are given by [2.3]

$$p_c = 22.0 \text{ MPa } (225\,\text{at}), \quad T_c = 374^\circ\,\text{C} = 647\,\text{K} \tag{2.5.15}$$

and the VAN DER WAALS *parameters* are [2.3]

$$\begin{aligned} a &= 555 \cdot 10^3\,\mathrm{N\,m^4\,kmol^{-2}} &&= \mathrm{kg\,m^5\,s^{-2}\,kmol^{-2}}, \\ b &= 0.0310\,\mathrm{m^3\,kmol^{-1}}, \quad C &&= 461\mathrm{J\,kg^{-1}}\,(\mathrm{m^2\,s^{-2}\,K^{-1}}). \end{aligned} \tag{2.5.16}$$

We now use (2.5.13) to generate a plot of the pressure volume relation of steam. We use the following Mathematica commands, which generate Fig. 2.1.

```
Clear[PP,GPP,T0,GT0,T1,GT1,T2,GT2]
$DefaultFont={''Courier-Bold'',10};
PP=Plot[8*0.6/(3*V-1)-3/V^2,{V,1,15}]
GPP=Graphics[PP];
T0=Text[A,{.8,0.05}];
GT0=Graphics[T0];
T1=Text[B,{4.,0.05}];
GT1=Graphics[T1];
T2=Text[C,{10.,0.2}];
```

```
GT2=Graphics[T2];
Show[GT0,GT1,GT2,GPP,Axes->True,Ticks->False,
AxesLabel->{''V'',''p''}]
```
(2.5.17)

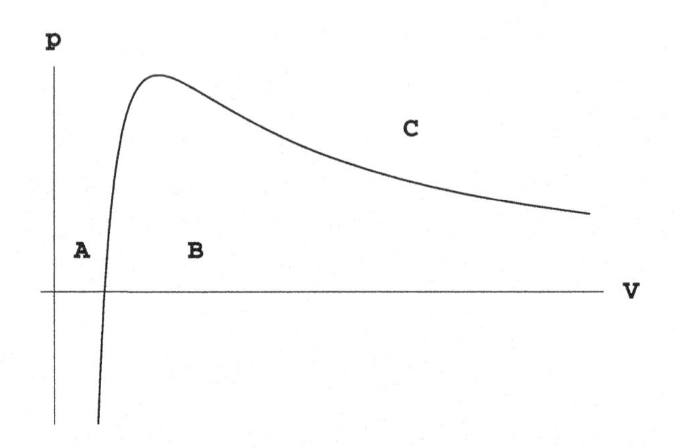


Fig. 2.1. Pressure volume relation of steam

In region A water is liquid, B is the two-phases region containing both water and steam. C is the gaseous region. The part of the curve separating A and B is called *saturated liquid line* and the part separating B and C is called *saturated vapor line.*

In the energy balance (2.4.2) the internal energy $U$ is still undefined. The first law of thermodynamics states that the sum of the heat importation $q$ and the exterior work $w$ done on the system must be equal to the change of internal energy: $\mathrm{d}U = q + w$. This implies that $\mathrm{d}U$ is a total differential. Since we have three state variables $p, V, T$ and one of those three may be eliminated using one equation of state like (2.5.2), (2.5.6) or (2.5.7), the *internal energy* may be written as $U(V,T)$, $U(p,T)$ or $U(V,p)$. Thus one may have

$$\mathrm{d}U = \left(\frac{\partial U}{\partial V}\right)_T \mathrm{d}V + \left(\frac{\partial U}{\partial T}\right)_V \mathrm{d}T, \tag{2.5.18}$$

$$\mathrm{d}U = \left(\frac{\partial U}{\partial p}\right)_T \mathrm{d}p + \left(\frac{\partial U}{\partial T}\right)_p \mathrm{d}T, \tag{2.5.19}$$

$$\mathrm{d}U = \left(\frac{\partial U}{\partial V}\right)_p \mathrm{d}V + \left(\frac{\partial U}{\partial p}\right)_V \mathrm{d}p. \tag{2.5.20}$$

The various partial derivatives have to be measured in experiments. For an isolated system kept on constant temperature ($\mathrm{d}T = 0$, *isothermal change of state* ) the overflow experiment by GAY-LUSSAC and other measurements may determine the partial differential quotients. For $q = 0$, $w = 0$, thus $\mathrm{d}U = 0$, equation (2.5.18) yields

$$\left(\frac{\partial U}{\partial V}\right)_T = 0 \tag{2.5.21}$$

for a perfect gas. If a thermodynamic system is kept at constant volume $\mathrm{d}V = 0$ (*isochoric change of state*) and heat $q$ is imported, then

$$q = \left(\frac{\partial U}{\partial T}\right)_V \mathrm{d}T = c_V \mathrm{d}T \tag{2.5.22}$$

defines the specific heat $c_V$ for constant volume. For a mol one uses $C_V$.

For the convenience of the reader we summarize some thermodynamic formulae. For a perfect gas one has

$$U(T) = \int_0^T c_V(T)\mathrm{d}T \approx c_V T + \text{const.} \tag{2.5.23}$$

Imported work is given by $p \cdot \mathrm{d}V$. We define *enthalpy* $H$ by

$$H(p,T) = U + pV \tag{2.5.24}$$

and obtain for an *isobaric change of state* $(\mathrm{d}p = 0)$

$$c_p = \left(\frac{\partial H}{\partial T}\right)_p = \left(\frac{\partial U}{\partial T}\right)_p + \left(\frac{\partial V}{\partial T}\right)_p \tag{2.5.25}$$

and

$$H(T) = \int_0^T c_p(T)\mathrm{d}T. \tag{2.5.26}$$

Now let us consider more general changes of state. We call *polytropic* a change of state if the imported heat $q$ may be described by a *polytropic specific heat c.* . If no heat is imported, $c = 0$, the change of state is called *adiabatic.*

$$q = c \cdot \mathrm{d}T = \mathrm{d}U + p\mathrm{d}V = \left[\left(\frac{\partial U}{\partial V}\right)_T + p\right] \mathrm{d}V + c_V \mathrm{d}T. \tag{2.5.27}$$

For an ideal gas $pV = RT$ one obtains (see problem 4)

$$\frac{c - c_p}{c - c_V} \frac{\mathrm{d}V}{V} + \frac{\mathrm{d}p}{p} = 0. \tag{2.5.28}$$

Integration yields the *polytropic change of state*

$$pV^n = \text{const}, \tag{2.5.29}$$

where

$$n = \frac{c - c_p}{c - c_V}. \qquad (2.5.30)$$

We now return to Eq. (2.5.27) and ask the question, if $q$ is an exact differential and if (2.5.27) can be integrated. We thus check if the mixed derivatives are equal:

$$\frac{\partial}{\partial T}\left[\left(\frac{\partial U}{\partial V}\right)_T + p\right] = \frac{\partial}{\partial V}\left(\frac{\partial U}{\partial T}\right)_V. \qquad (2.5.31)$$

We see that the term $(\partial p/\partial V)_V \neq 0$ indicates, that $q$ is not integrable: we need an *integrating factor* $J$. From

$$\frac{\partial}{\partial T}\left[\left(\frac{\partial U}{\partial V}\right)_T J + pJ\right] = \frac{\partial}{\partial V}\left[\left(\frac{\partial U}{\partial T}\right)_V J\right] \qquad (2.5.32)$$

we obtain $J = 1/T$. Then (2.5.27) takes the form

$$\frac{q}{T} = \mathrm{d}S = \frac{1}{T}\left(\frac{\partial U}{\partial V}\right)_T \mathrm{d}V + \frac{p}{T}\mathrm{d}V + \frac{c_v}{T}\mathrm{d}T \qquad (2.5.33)$$

which is integrable. The new function $S(T, V)$ or $S(p, T)$ or $S(V, p)$ is called *entropy.* It has the dimension $\mathrm{J\,K}^{-1}$. If (2.5.33) is valid, the thermodynamic process is called *reversible* (no dissipative effects like viscosity etc), but for $T\mathrm{d}S > q_{\mathrm{rev}}$ the process is called *irreversible.* $q_{\mathrm{rev}}$ designates heat import by a reversible process. Entropy increases and no potential flow (which is isentropic and polytropic) is possible.

For later use and for the convenience of the reader we now summarize the important five equations for the variables $u, \rho, p, T, U$, which will represent the basis for *tsunami* research ($D = 0$). We specialize to a one-dimensional problem (plane wave) and use the abbreviation $\mathrm{d}\,/\mathrm{d}t = \partial\,/\partial t + \vec{v}\nabla$ or

$$\frac{\mathrm{d}}{\mathrm{d}t} = \frac{\partial}{\partial t} + u\frac{\partial}{\partial x}. \qquad (2.5.34)$$

Then we have from (2.2.6) or (2.4.3) the *continuity equation*

$$\frac{\mathrm{d}\rho}{\mathrm{d}t} + \rho u_x = 0. \qquad (2.5.35)$$

From (2.3.13) the *equation of motion* takes the form

$$\rho\frac{\mathrm{d}u}{\mathrm{d}t} + p_x = \rho g + \eta u_{xx}. \qquad (2.5.36)$$

Instead of using the *energy balance* (2.4.2) for a flowing fluid for $Q = 0, \lambda =$ const we first derive the energy theorem from the equation of motion (2.5.36) by multiplication by $u$. We obtain the energy theorem for the purely mechanical terms in the form

$$\rho u \frac{\mathrm{d}u}{\mathrm{d}t} + u p_x = \rho u g + \eta u u_{xx}. \tag{2.5.37}$$

Now we have to add the thermodynamic terms $\rho(\mathrm{d}U/\mathrm{d}t)$, $\lambda T_{xx}$ and $p \operatorname{div} \vec{v} = p u_x$, see (2.4.2). $\Phi$ is now given by $\eta u_x^2$ (or $(4/3)u_x^2$). We thus have [2.18]

$$\rho \frac{\mathrm{d}}{\mathrm{d}t}\left(U + \frac{u^2}{2}\right) + \frac{\partial}{\partial x}(pu) = \eta \frac{\partial}{\partial x}(u u_x) + \lambda T_{xx} + \rho u g. \tag{2.5.38}$$

The fourth equation is given by the *state equation* of the fluid, may be (2.5.4) or (2.5.6) or (2.5.7). For a tsunami, represented by a fictive gas, we choose (2.5.4)

$$p = \rho C T. \tag{2.5.39}$$

The last equation stems from the definition of the *internal energy* $U$. From (2.5.23) we get

$$U(T) = \frac{c_V T C}{C} = c_V T, \tag{2.5.40}$$

where $C$ is defined by $R/M$ according to (2.5.5). According to (2.5.43), see problem 3, one has

$$C = c_p - c_V, \tag{2.5.41}$$

so that from (2.5.40)

$$U = \frac{c_V T C}{c_p - c_V} = \frac{CT}{(\kappa - 1)}. \tag{2.5.42}$$

A constant may be added and $\kappa = c_p/c_v$.

The tsunami wave equation which is a nonlinear partial differential equation of fifth order depending on two independent variables will be derived from the five equations (2.5.35), (2.5.36), (2.5.38), (2.5.39) and (2.5.42).

## Problems

1. Are the values (2.5.15), (2.5.16) and (2.5.10) given in the literature [2.3] consistent?

2. Derive some partial differential quotients for a gas defined by $T = pV/R$.

   Solutions:

$$\left(\frac{\partial U}{\partial p}\right)_V = \frac{c_V V}{R}; \quad \left(\frac{\partial U}{\partial V}\right)_p = \frac{c_V p}{R}; \quad \left(\frac{\partial U}{\partial p}\right)_T = 0.$$

3. Derive a formula for $c_p - c_V$.

   Solution:
$$c_p - c_V = \left[\left(\frac{\partial U}{\partial V}\right)_T + p\right]\left(\frac{\partial V}{\partial T}\right)_p. \tag{2.5.43}$$

   Specify for an ideal gas (2.5.4), observe (2.5.21) and receive $c_p - c_V = R/M = C$.

4. Derive (2.5.28).

   Solution for $T = pV/R$:
$$\left(\frac{\partial T}{\partial V}\right)_p = \frac{p}{R}; \quad q = c_V \mathrm{d}T + (c_p - c_V)p\mathrm{d}V/R = c\mathrm{d}T,$$
$$(c_p - c_V)p\mathrm{d}V = (c - c_V)(V\mathrm{d}p + p\mathrm{d}V),$$
$$RT = p\mathrm{d}V + V\mathrm{d}p,$$
$$c_p p\mathrm{d}V = cV\mathrm{d}p - c_V V\mathrm{d}p + cp\mathrm{d}V,$$

   which gives (2.5.28).

5. Show that *polytropic changes* $pV^n$ = const include isothermal, isobaric, isochoric and adiabatic changes of state.

   Solution:
$$\left.\begin{array}{lll} \text{isothermal} & n = 1, & c = \infty; \\ \text{isobaric} & n = 0, & c = c_p; \\ \text{isochoric} & n = \infty, & c = c_V; \\ \text{adiabatic} & n = \kappa = c_p/c_V, & c = 0, \\ \multicolumn{3}{l}{p/p_0 = (\rho/\rho_0)^\kappa,\ \rho/\rho_0 = (T/T_0)^{\frac{1}{(\kappa-1)}},\ p/p_0 = (T/T_0)^{\frac{\kappa}{(\kappa-1)}}.} \end{array}\right\} \tag{2.5.44}$$

6. Calculate the work $w = p\mathrm{d}V$ for isochoric, isothermal, isobaric and adiabatic changes of state of an ideal gas $p = RT/V$.

   Solution: isochoric: $w = 0$, isothermal: $w = RT(\ln V_2 - \ln V_1)$, isobaric: $p(V_2 - V_1)$, adiabatic: $c_V(T_2 - T_1)$.

7. Calculate $(\mathrm{d}p/\mathrm{d}\rho)$ for adiabatic, polytropic and isothermal changes of state of an ideal gas.

   Solution:
$$\frac{\mathrm{d}p}{\mathrm{d}\rho} = \frac{np}{\rho}, \quad \frac{\mathrm{d}p}{\mathrm{d}\rho} = pV. \tag{2.5.45}$$

Remark: $\sqrt{(\mathrm{d}p/\mathrm{d}\rho)} = c$ is a speed, for 0° C, 1 at, air one has 333 $\mathrm{m\,s^{-1}}$ for adiabatic and 280 $\mathrm{m\,s^{-1}}$ for isothermal change of state.

8. Calculate the adiabatic loss due to cooling of air (1 at, 293° K), if the air ascends to a height 2000 m (pressure 0.7 at).

   Solution: cooling 27.3° C.

9. Calculate the entropy $S$ for an ideal gas with constant specific heats $c_V, c_p$.

   Solutions:

$$\begin{aligned} S(T,V) &= c_V \ln T + R \ln V + \text{const} \\ S(T,p) &= c_p \ln T - R \ln p + \text{const} \\ S(p,V) &= c_V \ln (pV^\kappa) + \text{const}. \end{aligned} \tag{2.5.46}$$

   Thus adiabatic processes are isentropic.

10. Discuss the entropy production rate per unit volume [2.6]

$$\frac{\Phi}{T} + \frac{\lambda}{T^2}[(\nabla T)\cdot(\nabla T)] + \frac{1}{T}\delta q. \tag{2.5.47}$$

   Here $\Phi$ is the mechanical dissipation function and $\delta q$ is the rate of internal dissipation e.g. such as Joule heat of an electric conductive medium. Some authors [2.13] use

$$\rho T \frac{\mathrm{d}S}{\mathrm{d}t} = \eta u_x^2 + \frac{\partial}{\partial x}\lambda\left(\frac{\partial T}{\partial x}\right). \tag{2.5.48}$$

11. Calculate the *pressure density integral* (2.3.17) from $p_0$ to $p$ for adiabatic behaviour (2.5.44).

   Solution:

$$P = \frac{\kappa}{\kappa - 1}\left(\frac{p}{\rho} - \frac{p_0}{\rho_0}\right). \tag{2.5.49}$$

12. Using (2.4.1) in the form

$$\frac{v^2}{2} + P = 0,$$

   derive $v^2$ using (2.5.49).

   Solution (VENANT-WANTZEL *outflow formula*):

$$v^2 = \frac{2\kappa}{\kappa - 1}\frac{p_0}{\rho_0}\left[1 - \left(\frac{p}{p_0}\right)^{(\kappa-1)/\kappa}\right]. \tag{2.5.50}$$

If the outflow streams in the vacuum, $p = 0$ one obtains the maximum speed

$$v_m^2 = \frac{2\kappa p_0}{(\kappa - 1)\rho_0}. \tag{2.5.51}$$

This means that the total energy available is transformed into kinetic energy.

13. Equation (2.5.45) defines a speed. For adiabatic behavior of a gas, $c$ is the *sonic speed.* Using (2.5.50) and (2.5.45) derive a formula for the sonic speed.

    Solution:

$$c^2 = \frac{\kappa p_0}{\rho_0} - \frac{\kappa - 1}{2} v^2. \tag{2.5.52}$$

    Using the sonic speed $c_0$ of the gas at rest

$$c_0^2 = \kappa p_0 / \rho_0 \tag{2.5.53}$$

    one obtains

$$c^2 = c_0^2 - \frac{\kappa - 1}{2} v^2. \tag{2.5.54}$$

    Observe that the sonic speed in a gas flow depends on the velocity $v$ of the streaming gas.

14. Investigate (2.5.50) and plot the pressure-curve $p(v)$.
    Hints: Use (2.5.51) and designate $v/v_m$ by $x$ and $p/p_0$ by $y$. Use $\kappa = 1.401$ and $\kappa = 2$ (for later use in section 2.10). Plot $y(x)$ and investigate if this curve has a point of inflection:
    Solution:

$$y = (1 - x^2)^{\frac{\kappa}{\kappa - 1}} \tag{2.5.55}$$

    and use the Mathematica command `k=1.401;k/(k-1);`
```
y[x_]=(1-x^2)^(k/(k-1));
Plot[y[x],{x,0.001,1.}]
f[x]=InputForm[D[y[x],{x,2}]]
```
    The inflection point may be found by plotting $f[x]$ from $x = 0.1 - 0.8$.

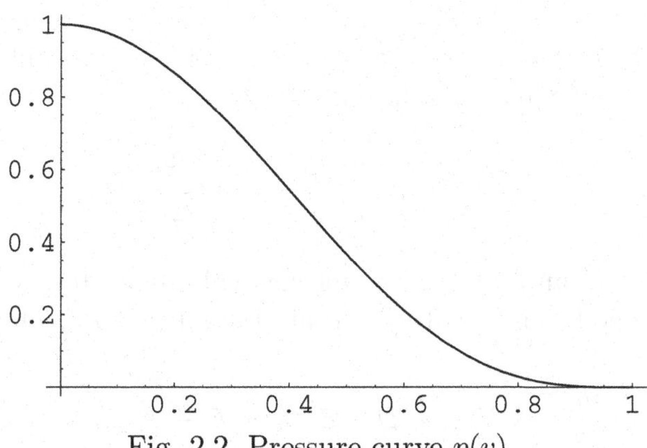


Fig. 2.2. Pressure-curve $p(v)$

## 2.6 Vorticity theorems

Since the rotary flow in hurricanes is the source of their destruction power it is important to investigate rotational motions and vortices in fluids. For this purpose we first derive the CROCCO *theorem* connecting vorticity and entropy. *Vorticity* is expressed by curl $\vec{v}$.

Neglecting external heat import $\partial Q/\partial t$ and taking the gravity potential $V$ into account, the energy balance (2.4.2) may be written

$$\rho \frac{\mathrm{d}U}{\mathrm{d}t} + p\nabla\vec{v} = \Phi + \lambda\Delta T = q = T\mathrm{d}S. \tag{2.6.1}$$

Here we used (2.5.33) and (2.5.34). With the help of (2.4.4) the Eq. (2.6.1) may be transformed into

$$\rho \frac{\mathrm{d}U}{\mathrm{d}t} + \frac{p}{\rho}D + p\rho\frac{\mathrm{d}(1/\rho)}{\mathrm{d}t} = \Phi + \lambda\Delta T = q = TdS. \tag{2.6.2}$$

Thus one has the *first law* of thermodynamics in the form

$$\left.\begin{aligned} \mathrm{d}U + \frac{p}{\rho^2}D + p\frac{\mathrm{d}(1/\rho)}{\mathrm{d}t} &= TdS/\rho \\ \nabla U + \frac{p}{\rho^2}D + p\nabla\frac{1}{\rho} &= T\nabla S/\rho, \end{aligned}\right\} \tag{2.6.3}$$

since $\mathrm{d}U = \mathrm{d}\vec{x}\cdot\nabla U$. Now we would like to consider the mechanical terms. The equation of motion (2.3.14) reads

$$\frac{\partial\vec{v}}{\partial t} + \nabla\frac{\vec{v}^{\,2}}{2} - \vec{v}\,\mathrm{curl}\,\vec{v} = -\frac{1}{\rho}\nabla p - \nabla V + \frac{1}{\rho}\vec{F}. \tag{2.6.4}$$

Here $\vec{F}$ are the friction forces due to the viscosity of the fluid. Now we define a *generalized enthalpy* by modifying (2.5.24)

$$H = U + \frac{p}{\rho} + V + \frac{\vec{v}^2}{2}. \tag{2.6.5}$$

By using this $H$ and adding the mechanical energy terms (2.6.4) and the thermodynamic terms (2.6.3) we obtain the CROCCO *theorem:*

$$\frac{\partial \vec{v}}{\partial t} + \nabla H + \frac{p}{\rho^2} D - \vec{v} \times \operatorname{curl} \vec{v} - \frac{\vec{F}}{\rho} = T \nabla S / \rho. \tag{2.6.6}$$

The theorem connects vorticity curl $\vec{v}$ with an increase of entropy [2.11]. The heat conduction term does not appear explicitly. We obtain the following conclusions:

For an isentropic flow $S =$ const, $q = 0$ (adiabatic change of state, there must be no heat conduction, no other dissipative (viscous) effects) the flow must be irrotational ($\operatorname{curl} \vec{v} = 0$) or a BELTRAMI *flow* $\vec{v} \times \operatorname{curl} \vec{v} = 0$, the flow must be stationary ($\partial \vec{v} / \partial t = 0$) and no source $D$ can be present. A potential flow ($\vec{v} = \nabla \varphi$) is always isentropic and polytropic.

In view of hurricanes we now investigate the possible mechanisms of vortex generation. First we define *circulation* by

$$\Gamma = \oint \vec{v} \mathrm{d}\vec{s} = \int_F \operatorname{curl} \vec{v} \mathrm{d}\vec{f}. \tag{2.6.7}$$

Here $\mathrm{d}\vec{s}$ is a line element and STOKES *theorem* has been used. Then the rate of change for $\Gamma$ is given by

$$\frac{\mathrm{d}\Gamma}{\mathrm{d}t} = \frac{\mathrm{d}}{\mathrm{d}t} \oint \vec{v} \mathrm{d}\vec{s} = \oint \frac{\mathrm{d}\vec{v}}{\mathrm{d}t} \mathrm{d}\vec{s} + \oint \vec{v} \mathrm{d}\frac{\mathrm{d}\vec{s}}{\mathrm{d}t} = \oint \frac{\mathrm{d}\vec{v}}{\mathrm{d}t} \mathrm{d}\vec{s} + \oint \vec{v} \mathrm{d}\vec{v}. \tag{2.6.8}$$

Since $\oint \vec{v} \mathrm{d}\vec{v} = \oint \mathrm{d}(\vec{v}^2/2) = \oint \nabla(\vec{v}^2/2) = 0$, we have only one nonvanishing term. Using the dissipationless NEWTON's *law* (2.3.13) we obtain (for $\eta = 0$)

$$\frac{\mathrm{d}\Gamma}{\mathrm{d}t} = \oint \frac{\mathrm{d}\vec{v}}{\mathrm{d}t} \mathrm{d}\vec{s} = \oint \left( -\frac{\nabla p}{\rho} - \vec{g} \right) \mathrm{d}\vec{s}. \tag{2.6.9}$$

Since the "*solenoidal term*" $\nabla p / \rho$ may be replaced by $\nabla P$ according to (2.3.17), this means, if a polytropic change of state occurs (2.5.44), then $\oint (\nabla P) = 0$. Since gravity $\vec{g}$ may be represented by a scalar potential, also the term $\oint \vec{g} \mathrm{d}\vec{s}$ vanishes. This proves THOMSONs *circulation theorem:*

If the exterior forces $\vec{g}$ are conservative and if the fluid is polytropic and thus inviscid the circulation $\Gamma$ is conserved. (Circulation along a closed curve does not change in time). It is however no longer constant, if the force $\vec{g}\,'$ as e.g. the CORIOLIS *force* has to be taken into account. A *hurricane* which is a circular vortex gains circulation from the CORIOLIS force $\vec{g}\,'$, see later. If heat is added to the system then it is no longer polytropic and convection destroys the conservation of the circulation.

Vorticity of a fluid is described by the *vorticity vector* $\vec{w}$

$$\vec{w} = (\operatorname{curl}\vec{v})/2. \tag{2.6.10}$$

Since $(\mathrm{d}\Gamma/\mathrm{d}t) = 0$ in an inertial coordinate system (no CORIOLIS-, no centrifugal forces) one has

$$\frac{\mathrm{d}}{\mathrm{d}t}\oint \vec{v}\mathrm{d}\vec{s} = 0 = \frac{\mathrm{d}}{\mathrm{d}t}\int_F 2\vec{w}\mathrm{d}\vec{f} = 0. \tag{2.6.11}$$

This means that the *vortex flow* $\int \vec{w}\mathrm{d}\vec{f}$ is conserved. HELMHOLTZ stated this as follows: No fluid particles can have a rotation if it did not originally rotate (HELMHOLTZ *first vorticity theorem*). Since $\vec{w}$ is proportional to the angular frequency $\omega$ of rotation, one may say that $\omega \cdot F = \text{const}$: a smaller cross section $F$ rotates faster. This is valid in dissipative free systems. If viscosity is taken into account, then we have to start from (2.3.13), (2.3.18)

$$\begin{aligned}\frac{\mathrm{d}\vec{v}}{\mathrm{d}t} = \frac{\partial\vec{v}}{\partial t} + (\vec{v}\nabla)\cdot\vec{v} = \frac{\partial\vec{v}}{\partial t} + \nabla\frac{\vec{v}^{\,2}}{2} - \vec{v}\times\operatorname{curl}\vec{v} = \\ -\frac{\nabla p}{\rho} - \nabla V + \frac{\eta}{\rho}\Delta\vec{v}.\end{aligned} \tag{2.6.12}$$

For an incompressible fluid ($\rho = \text{const}$) curl $\nabla p = 0$, and for conservative forces curl $\nabla V = 0$. Furthermore, curl grad $(v^2/2) = 0$. Then the application of curl on (2.6.12) yields ($\mu = \eta/\rho$)

$$\frac{\mathrm{d}\vec{w}}{\mathrm{d}t} = \frac{\partial\vec{w}}{\partial t} + (\vec{v}\nabla)\vec{w} = \mu\Delta\vec{w} + (\vec{w}\nabla)\vec{v} + \operatorname{curl}\vec{g}\,'. \tag{2.6.13}$$

This equation (*vorticity transfer equation*) describes vorticity transport in incompressible viscous fluid in a non-inertial coordinate system. If $\vec{g}\,' = 0$, the equation states that viscous effects do NOT generate new vortices. One may also say that fluid particles which are at any time part of a vortex filament always belong to that same vortex filament (HELMHOLTZ *second vorticity theorem*). An analysis of (2.6.13) shows that viscosity damps and non-conservative forces induce vortices.

**Problems**

1. Derive (2.6.13).

2. The one-dimensional vorticity transfer equation for $w(x,t)$ reads
$$\frac{\partial w}{\partial t} = \frac{\mathrm{d}w}{\mathrm{d}t} = \mu \frac{\partial^2 w}{\partial x^2}. \tag{2.6.14}$$
Solve this equation.

   Solution: $\dfrac{\text{const}}{\sqrt{4\mu t}} \exp\left(-\dfrac{x^2}{4\mu t}\right)$.

3. A polytropic atmosphere may be defined by
$$p = p_0 \left(\frac{\rho}{\rho_0}\right)^n, \quad \rho = \rho_0 \left(\frac{p}{p_0}\right)^{1/n}. \tag{2.6.15}$$
For the static equilibrium on Earth assume
$$gz + \frac{p_0^{1/n}}{\rho_0} \cdot \frac{n}{n-1} p^{1-1/n} = \text{const}.$$
Assume $p(h) = 0$, where $h$ is the height of the atmosphere. Calculate $p(z)$.

   Solution:
$$p(z) = p_0(1 - z/h)^{n/(n-1)}, \quad \rho(z) = \rho_0(1 - z/h)^{n/(n+1)}$$
($p_0 = 1\,\text{at} = 1033.00981\,\text{g' cm}^{-1}\,\text{s}^{-2}$, $\rho = 0.0013\,\text{g cm}^{-3}$, $\text{g} = 981\,\text{cm s}^{-2}$.)
The usual barometric formula is however
$$p = p_0 \exp(-Mgz/RT). \tag{2.6.16}$$
Here $p_0$ is the pressure at $z = 0$, $M$ is the molecular weight of air and $R$ is the universal gas constant. The formula is valid for an isothermal gas of constant temperature.

## *2.7 Potential flow in incompressible fluids*

For an incompressible fluid the basic equations are now modified due to $\rho =$ const. With some exceptions one may assume that water is practically incompressible. Thus the continuity equation (2.2.4) now reads

$$\operatorname{div} \vec{v} = D/\rho \tag{2.7.1}$$

and the equation of motion (2.3.13) yields

$$\frac{\partial \vec{v}}{\partial t} + (\vec{v}\,\nabla)\vec{v} = -\frac{\nabla p}{\rho} + \vec{g} + \nu\Delta\vec{v}. \qquad (2.7.2)$$

The last rhs term in this equation describes dissipation due to viscosity. Since any dissipative effect increases entropy and since we want to investigate potential flow, one has curl $\vec{v} = 0$ and the CROCCO *theorem* (2.6.6) takes the form

$$\frac{\partial \vec{v}}{\partial t} + \nabla H + \frac{p}{\rho^2} D = 0. \qquad (2.7.3)$$

Neglecting exterior forces $\vec{g}, \vec{g}\,'$ and using the definition for $H$ (2.6.5) one obtains

$$\rho\frac{\partial \vec{v}}{\partial t} + \rho\nabla U + \rho\nabla\frac{\vec{v}^{\,2}}{2} + \nabla p + \frac{pD}{\rho} = 0, \qquad (2.7.4)$$

which may be called a *time dependent* BERNOULLI *equation*, compare (2.3.20). In view of the coming sections we will investigate two special cases:

1. steady potential flow with $D = 0$ in two dimensions,

2. time-dependent flow.

For time-independent flow with $D = 0$, one may forget (2.7.2)–(2.7.4) and start with

$$\vec{v} = \nabla\varphi. \qquad (2.7.5)$$

Inserting into (2.7.1) we obtain the elliptic *potential equation* (LAPLACE *equation*)

$$\Delta\varphi = 0. \qquad (2.7.6)$$

If $D \neq 0$, one obtains a POISSON *equation*

$$\Delta\varphi = D/\rho. \qquad (2.7.7)$$

This is an inhomogeneous and elliptic equation and may be reduced to (2.7.6), see p 141 in [1.1]. In two dimensions (2.7.6) is reduced to *conformal mapping.* For conformal mapping in three dimensions see p 165 in [1.1]. A special solution of (2.7.6) is of importance with respect to hurricanes: the two-dimensional *circular vortex.*

In conformal mapping the two-dimensional LAPLACE equation (2.7.6) is usually replaced by the CAUCHY-RIEMANN *equations*

$$v_x = -\frac{\partial \varphi}{\partial x} = -\frac{\partial \psi}{\partial y}, \quad v_y = -\frac{\partial \varphi}{\partial y} = \frac{\partial \psi}{\partial x}. \qquad (2.7.8)$$

Now introducing a complexe coordinate $z$ and a complexe potential $\zeta$

$$z = x + iy = r\exp(i\chi), \quad \zeta(x,y) = \varphi(x,y) + i\psi(x,y) \tag{2.7.9}$$

one may again define circulation

$$\Gamma = \oint \zeta(z)\mathrm{d}z = \oint (\varphi(x,y) + i\psi(x,y))(\mathrm{d}x + i\mathrm{d}y). \tag{2.7.10}$$

If no singularities are present within the domain, then one has the CAUCHY *integral* formula

$$\oint \zeta(z) \cdot \mathrm{d}z = 0. \tag{2.7.11}$$

If the integration path in the complexe plane $x, iy$ encloses a singularity, then

$$\oint \zeta(z)\mathrm{d}z = 2\pi i \mathrm{Res},$$

where Res is the so-called residuum of the singularity. An example may be given by

$$\zeta(z) = -\frac{i\Gamma}{2\pi}\ln z = \frac{-i\Gamma}{2\pi}\ln r + \frac{\Gamma}{2\pi}\arctan\frac{y}{x}, \tag{2.7.12}$$

where $\Gamma$ is a constant. Splitting up into real and imaginary parts one has

$$\varphi = \frac{\Gamma}{\pi}\arctan\frac{y}{x}, \quad \psi = -\frac{\Gamma}{2\pi}\ln\sqrt{x^2+y^2}. \tag{2.7.13}$$

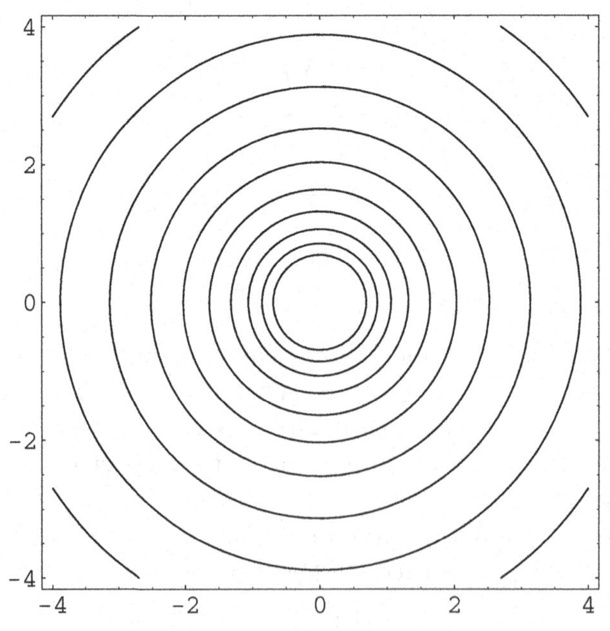


Fig. 2.3. Circular vortex

Mathematica may again help. To generate Fig. 2.3 one uses the commands

```
Clear[psi];psi[x_,y_]=-Log[Sqrt[x^2+y^2]];
C1=ContourPlot[psi[x,y],{x,-4.,4.},{y,-4.,4.},                (2.7.14)
ContourShading->False,ContourSmoothing->2,
PlotPoints->80]
```

Next we consider time dependent potential solutions. From (2.7.3), (2.7.5) we have for $D = 0$

$$\nabla\left(\frac{\partial\varphi}{\partial t} + \frac{v^2}{2} + \frac{p}{\rho} + gz\right) = 0. \tag{2.7.15}$$

Here we assumed constant temperature $(\nabla U = 0), \vec{v} = \nabla\varphi$ and gravity in the $z$-direction as external force. Putting $p_0/\rho_0$ into an integration constant $C$, we may write

$$\Delta\varphi = 0, \tag{2.7.16}$$

$$\frac{\partial\varphi}{\partial t} + \frac{(\nabla\varphi)^2}{2} + gz = C. \tag{2.7.17}$$

Here we assumed that $z = 0$ indicates a free water surface, for instance of a lake or ocean, with constant depth $z = -h$. At the ocean bottom we have to assume that the vertical motion vanishes. Thus we have the *bottom boundary condition*

$$\left(\frac{\partial\varphi}{\partial z}\right)_{z=-h} = 0. \tag{2.7.18}$$

On the other hand, at the free surface $z = 0$ the pressure $p$ is constant so that

$$\left(\frac{\partial\varphi}{\partial t}\right)_{z=0} + (\nabla\varphi)^2 = C. \tag{2.7.19}$$

A liquid particle at the location $z$ will have a vertical velocity given by

$$\frac{\mathrm{d}z}{\mathrm{d}t} = v_z = \frac{\partial\varphi}{\partial z}. \tag{2.7.20}$$

These equations will play an important role in the investigation of water waves in chapter 3.

## Problems

1. Derive (2.7.6) from (2.7.8).

2. Prove that (2.7.13) is a solution of (2.7.8).

3. Neglect $(\nabla\varphi)^2$ in (2.7.17) and derive a wave equation for $\varphi(x,y,z,t)$.

   Solution:
$$\frac{\partial^2\varphi}{\partial t^2} = \pm g\frac{\partial\varphi}{\partial z}. \tag{2.7.21}$$

4. Solve (2.7.18) by the setup
$$\varphi(x,y,z,t) = \psi(x,y)\cdot U(z)\cdot \exp(i\omega t). \tag{2.7.22}$$

5. Describe the two-dimensional steady incompressible potential flow around a cylinder by conformal mapping using Mathematica.

   Solution [1.1]

```
giza1[z_]=z/(2*V0)+Sqrt[z^2/(4*V0^2)-R^2];
<<Graphics`ComplexMap`
echo $Packages
Clear[R,V0,P1];R=1.;V0=1.;
P1=CartesianMap[giza1,{-6.,6.},{-6.,6.}];
Clear[giza2,P2];giza2[z_]=z/(2*V0)-Sqrt[z^2/(4*V0^2)-R^2];
P2=CartesianMap[giza2,{-6.,6.},{-6.,6.}];
Show[P1,P2,PlotRange->All]
```

## *2.8 Potential flow in compressible fluids*

Due to the *equivalence theorem* stating that solutions of compressible fluid equations are equivalent to solutions of nonlinear water wave equations, we have a need to investigate gasdynamics. We assume $D = 0, U =$ const, $\operatorname{curl}\vec{v} = 0, \vec{v} = -\nabla\varphi$. We first rewrite the *continuity equation* (2.2.4) in the form
$$\frac{1}{\rho}\frac{\partial\rho}{\partial t} - \Delta\varphi - \nabla\varphi\cdot\frac{\nabla\rho}{\rho} = 0. \tag{2.8.1}$$

The *time dependent* BERNOULLI *equation* (2.7.4) reads now
$$-\nabla\frac{\partial\varphi}{\partial t} + \frac{1}{2}\nabla(\nabla\varphi)^2 + \nabla P = 0. \tag{2.8.2}$$

$P$ is defined by (2.3.17), (2.5.49). No heat exchange or dissipative effects occur in a potential flow, it is adiabatic. Thus (2.5.45) may read
$$c^2\nabla\rho = \nabla p; \quad \nabla P = c^2\cdot\frac{\nabla\rho}{\rho}; \quad \mathrm{d}P = c^2\frac{\mathrm{d}\rho}{\rho}. \tag{2.8.3}$$

Integration over space of (2.8.2) yields

$$-\frac{\partial \varphi}{\partial t} + \frac{1}{2}(\nabla\varphi)^2 + P = \text{const} = F(t) \to 0. \tag{2.8.4}$$

Derivation with respect to time gives

$$-\frac{\partial^2 \varphi}{\partial t^2} + \frac{1}{2}\frac{\partial}{\partial t}(\nabla\varphi)^2 + \frac{\partial P}{\partial t} = 0. \tag{2.8.5}$$

Using (2.8.3) one has

$$\frac{\partial P}{\partial t} = \frac{c^2}{\rho}\frac{\partial \rho}{\partial t}, \tag{2.8.6}$$

so that (2.8.5) reads

$$\frac{1}{\rho}\frac{\partial \rho}{\partial t} = \frac{1}{c^2}\frac{\partial P}{\partial t} = \frac{1}{c^2}\left[\frac{\partial^2 \varphi}{\partial t^2} - \frac{1}{2}\frac{\partial}{\partial t}(\nabla\varphi)^2\right]. \tag{2.8.7}$$

Now we can insert $\frac{1}{\rho}\frac{\partial \rho}{\partial t}$ and from (2.8.2) and (2.8.3) we insert

$$\frac{\nabla\rho}{\rho} = \frac{1}{c^2}\nabla P = \frac{1}{c^2}\left[\nabla\frac{\partial \varphi}{\partial t} - \frac{1}{2}\nabla(\nabla\varphi)^2\right] \tag{2.8.8}$$

into the continuity equation (2.8.1) obtaining the *potential equation* for compressible fluids

$$\frac{1}{c^2}\left[\frac{\partial^2 \varphi}{\partial t^2} - \frac{1}{2}\frac{\partial}{\partial t}(\nabla\varphi)^2\right] - \Delta\varphi - \frac{\nabla\varphi}{c^2}\left[\nabla\frac{\partial \varphi}{\partial t} - \frac{1}{2}\nabla(\nabla\varphi)^2\right] = 0. \tag{2.8.9}$$

We will treat *tsunamis* as nonlinear water waves. Thus we are interested into the potential $\varphi(x,t)$. For small wave amplitudes the nonlinear terms $(\nabla\varphi)^2$ may be neglected and (2.8.9) yields the hyperbolic acoustic wave equation

$$\frac{1}{c^2}\frac{\partial^2 \varphi}{\partial t^2} = \Delta\varphi. \tag{2.8.10}$$

If the fluid is incompressible, $\rho =$ const, then according to (2.8.3) $\mathrm{d}\rho/\mathrm{d}p = 1/c^2$ or $c \to \infty$. Elastic waves propagate faster in nearly incompressible media far faster than in water. For steel one has $c = 5100$ m s$^{-1}$, for water at 0° one has 1403 m $s^{-1}$, at 20° however 1483 m $s^{-1}$ and for air 332 m $s^{-1}$ at 0° C. Thus we may use $\Delta\varphi = 0$. The equation for $\varphi(x,t)$ is received from (2.8.9). It reads

$$F\left(x, t, \varphi_x, \varphi_t, \varphi_{xx}, \varphi_{xt}, \varphi_{tt}\right) = \varphi_{xx}\left(c^2 - \varphi_x^2\right) - 2\varphi_x\varphi_{xt} - \varphi_{tt} = 0, \quad (2.8.11)$$

where now

$$c^2 = c_0^2 - \frac{(\kappa - 1)}{2}\varphi_x^2 - (\kappa - 1)\varphi_t, \quad (2.8.12)$$

compare (2.5.52) and problem 1. Now we use the mathematics developed in section (1.2) and (1.4) to classify the new nonlinear partial differential equation (2.8.11). From (1.4.29) and (2.8.11) we have

$$a = (c^2 - \varphi_x^2), \quad b = \varphi_x, \quad c = -1, \quad (2.8.13)$$

where $c$ should not be confused with the sonic speed $c$ defined by (2.8.12). If we insert (2.8.13) into the criterion (1.2.7), we find $\sqrt{b^2 - ac} > 0$ so that we conclude that (2.8.11) is a hyperbolic nonlinear partial differential equation. This is true for $c > \varphi_x = v$. Since in this case the sonic speed is larger than the flow speed, one speaks of *supersonic flow.* If $c < v$ one uses the term *subsonic flow* and (2.8.11) is elliptic and the characteristics (2.8.14) become complex. The characteristics

$$w = \frac{\mathrm{d}x}{\mathrm{d}t} = \varphi_x \pm \sqrt{c_0^2 - \frac{(\kappa - 1)}{2}\varphi_x^2 - (\kappa - 1)\varphi_t} = u \pm c(u) \quad (2.8.14)$$

depend on the still unknown solution itself and are hence of no value. If our equation would not be nonlinear, but linear, the characteristics would be of great value. Actually we can exactly linearize (2.8.11) using a LEGENDRE *transformation.* For this purpose we define [1.1]

$$q = -\varphi_t, \; u = -\varphi_x, \; -u_x = \varphi_{xx}, \; -q_t = \varphi_{tt}. \quad (2.8.15)$$

Then

$$\mathrm{d}\varphi = \varphi_x \mathrm{d}x + \varphi_t \mathrm{d}t = -u\mathrm{d}x - q\mathrm{d}y. \quad (2.8.16)$$

Next we define a new potential

$$\Psi(u, q) = ux + qt + \varphi(x, t), \quad (2.8.17)$$

$$\mathrm{d}\Psi = \Psi_u \mathrm{d}u + \Psi_q \mathrm{d}q = x\mathrm{d}u + t\mathrm{d}q, \tag{2.8.18}$$

where

$$\Psi_u = x, \quad \Psi_q = t, \quad \Psi_{uu} = x_u, \quad \Psi_{uq} = t_q. \tag{2.8.19}$$

We now solve for $\varphi$

$$\varphi_{xx} = -\Psi_{qq}/D, \quad \varphi_{xt} = \Psi_{qv}/D, \quad \varphi_{tt} = -\Psi_{uu}/D, \tag{2.8.20}$$

where

$$D = \begin{vmatrix} \Psi_{uu} & \Psi_{uq} \\ \Psi_{uq} & \Psi_{qq} \end{vmatrix}. \tag{2.8.21}$$

Inserting into (2.8.11) we obtain the new *linear* potential equation

$$F = \Psi_{qq}\left(c^2 - u^2\right) + 2u\Psi_{uq} - \Psi_{uu} = 0, \tag{2.8.22}$$

$$c^2 = c_0^2 - \frac{(\kappa - 1)}{2} u^2 + (\kappa - 1)q. \tag{2.8.23}$$

Now we consider again the criterion (1.2.7). From (2.8.22) we recognize that (2.8.22) is of hyperbolic type and the characteristics in the $u, q$ plane are

$$\frac{\mathrm{d}q}{\mathrm{d}u} = (u \pm c). \tag{2.8.24}$$

This is independent from $q$. The *wave propagation speed* $w$ depends on $u$ according to (2.8.14). This means that a wave propagates faster in relation to the speed of flow velocity $u$. Therefor a pressure wave amplitude will steepen up during propagation. This continual steepening up of the wave front can no longer be described by single-valued functions of location. One has to introduce a discontinuity into the flow to get over this difficulty. This discontinuity is called a *shock wave*, across which the flow variables change discontinuously. The discontinuity vanishes if the basic fluid equations are modified to contain viscosity and heat conduction. Figures 2.4 and 2.5 depict the steepening up of a large amplitude gas wave.

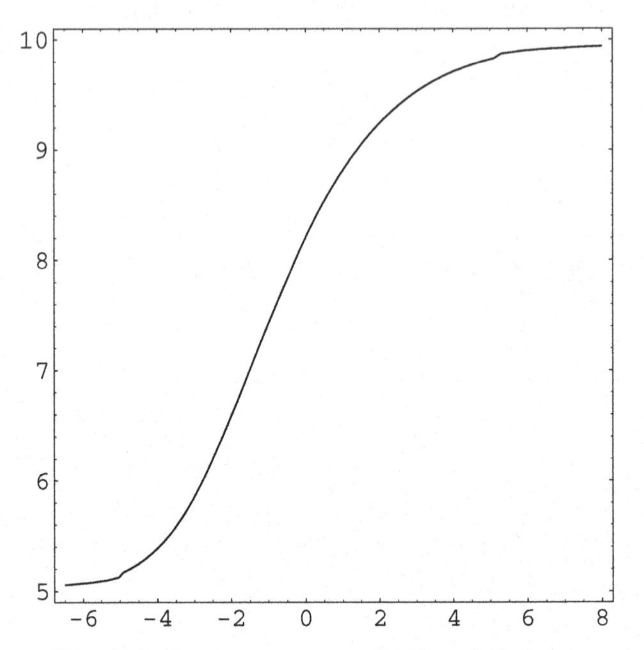


Fig. 2.4. Large wave travelling to left at time $t_0$

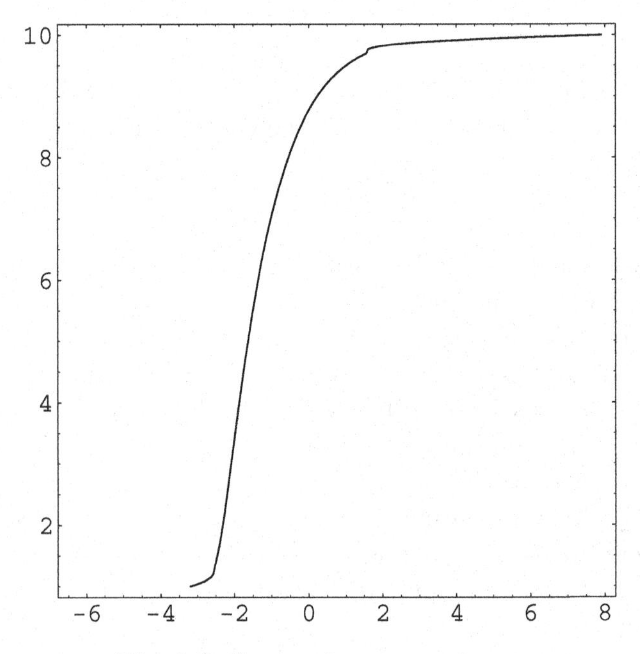


Fig. 2.5. Steepening up at time $t > t_0$

The high-pressure region of the travelling wave progresses more quickly than the low-pressure region. Then the former overtakes the latter, and the pressure rise associated with the wave then occurs discontinuously through a very (even infinitely) steep front. Thus the wave becomes a shock wave. The flowing gas is rapidly compressed by its passage through the shock. If steepening of the shock would continue, topple over would appear and an unphysical double-valued function would describe the phenomenon. Anyway, conservation of mass, momentum and energy must be valid over a shock. If subscript 1 designates the situation in front of the shock and subscript 2 after the shock, then conservation of mass is given by

$$\rho_1 u_1 = \rho_2 u_2, \tag{2.8.25}$$

momentum conservation is expressed by

$$p_1 + \rho_1 u_1^2 = p_2 + \rho_2 u_2^2 \tag{2.8.26}$$

and energy conservation yields

$$U_1 + \frac{u_1^2}{2} + \frac{p_1}{\rho_1} = U_2 + \frac{u_2^2}{2} + \frac{p_2}{\rho_2}. \tag{2.8.27}$$

Using

$$U = c_V T \tag{2.8.28}$$

and

$$\frac{p}{\rho} = RT = (c_p - c_V)T, \quad \kappa = c_p / c_V \tag{2.8.29}$$

one can derive the HUGENIOT *state equation*

$$\frac{p_2}{p_1} - \frac{\rho_2}{\rho_1} = \frac{\kappa - 1}{2}\left(1 + \frac{p_2}{p_1}\right)\left(\frac{\rho_2}{\rho_1} - 1\right). \tag{2.8.30}$$

$p_2/p_1$ as a function of $\rho_2/\rho_1$ is steeper than the adiabatic curve and reaches

$$\frac{p_2}{p_1} \to \infty \quad \text{at} \quad \frac{\rho_2}{\rho_1} = \frac{\kappa+1}{\kappa-1} \quad \left(\text{for air } \frac{\rho_2}{\rho_1} \approx 6\right).$$

It is clear that entropy is increased in a shock.

Behind a shock wave a *rarefaction wave* is to be expected due to (2.8.25). This again might excite a second but weaker large amplitude wave.

Equations (2.8.25)–(2.8.27) are valid only for a stationary flow. For a time-dependent flow, these equations have to be modified. If we designate by $w$ the propagation speed of a *travelling shock wave*, then (2.8.25)–(2.8.27) assume the following form [2.12], [2.19]

$$(w - u_1)\,\rho_1 = (w - u_2)\,\rho_2, \tag{2.8.31}$$

$$(w - u_1)^2\,\rho_1 + p_1 = (w - u_2)^2\,\rho_2 + p_2, \tag{2.8.32}$$

$$\frac{a_1^2}{\kappa - 1} + \frac{(w - u_1)^2}{2} = \frac{a_2^2}{\kappa - 1} + \frac{(w - u_2)^2}{2}. \tag{2.8.33}$$

Here we had used the expressions

$$a_1^2 = \frac{\kappa p_1}{\rho_1}, \qquad a_2^2 = \frac{\kappa p_2}{\rho_2} \tag{2.8.34}$$

for the sonic speeds $a_1$ and $a_2$. The solution for $w$ yields (see problem 4)

$$w = \frac{a_1^2 - a_2^2}{\kappa - 1} \cdot \frac{1}{u_1 - u_2} + \frac{u_1 + u_2}{2}. \tag{2.8.35}$$

If one inserts this result into (2.8.31) and (2.8.32) one obtains the *unsteady shock polar* [2.12]

$$u_2 = u_1 \pm \sqrt{\frac{1}{\kappa}\left[-\left(a_1^2 + a_2^2\right) + \sqrt{\left(a_1^2 + a_2^2\right)^2 - \frac{4\kappa^2}{\kappa - 1}\left[\frac{1}{\kappa} - \frac{1}{\kappa - 1}\right]\left(a_1^2 - a_2^2\right)^2}\right]} \tag{2.8.36}$$

All this mathematics is of no interest for hurricanes, but it allows insight into the mechanism of large amplitude water waves like solitons, tsunamis, water jumps, swells, water hammer etc.

## Problems

1. Use (2.5.48), (2.5.51) and integrate (2.8.8) to derive (2.8.12).

   Solution:

$$\frac{1}{c^2}P = \frac{1}{c^2}\left[\frac{\partial \varphi}{\partial t} - \frac{1}{2}(\nabla\rho)^2\right]; \quad P = \frac{1}{\kappa - 1}\left(\frac{\kappa p}{\rho} - \frac{\kappa p_0}{\rho_0}\right); \quad \frac{\kappa p_0}{\rho_0} = c_0^2.$$

2. Derive (2.8.24) and prove that (2.8.22) is hyperbolic.

   Solution: From (1.4.29) and (2.8.22) one obtains

$$a = \left(c^2 - u^2\right), \quad b = -u, c = -1; \quad b^2 - ac > 0.$$

3. Derive (2.8.30).

4. Derive (2.8.35). Hints: divide (2.8.32) by (2.8.31) and use (2.8.34), then use (2.8.33).

5. Derive (2.8.36).

6. Is the linear equation (2.8.22) separable into two ordinary differential equations by $\Psi(u,q) = U(u)Q(q)$?

   Solution: No.

7. What do the following equations describe [2.13]?

   Solution:

$$\left.\begin{aligned}
&\frac{\partial \rho}{\partial t} + \frac{\partial}{\partial x}(\rho u) = 0, \\
&\frac{\partial}{\partial t}(\rho u) + \frac{\partial}{\partial x}\left(\rho u^2 + p - \frac{4}{3}\eta u_x\right) = 0, \\
&\frac{\partial}{\partial t}\left(\rho U + \rho\frac{u^2}{2}\right) + \frac{\partial}{\partial x}\left(\rho u U + \rho\frac{u^2}{2}u + u\left(p - \frac{4}{3}\eta u_x\right) - \lambda T_x\right) = 0, \\
&\rho T\left(\frac{\partial S}{\partial t} + u\frac{\partial S}{\partial x}\right) = \frac{4}{3}\eta u_x^2 + \frac{\partial}{\partial x}(x, T_x) = 0
\end{aligned}\right\} \tag{2.8.37}$$

8. Derive the characteristics of an unsteady one-dimensional compressible flow described by

$$\left.\begin{aligned}
&\rho_t + u\rho_x + \rho u_x = 0 \\
&u_t + uu_x + \frac{1}{\rho}p_x = 0, \\
&S_t + uS_x = 0, \\
&p = p(\rho, S), \quad p_x = p_\rho \rho_S + p_S S_x
\end{aligned}\right\} \tag{2.8.38}$$

   Hint: Write the system in the form

$$U_t + AU_x = 0, \tag{2.8.39}$$

where

$$U = \begin{pmatrix} \rho \\ u \\ S \end{pmatrix}, \quad A = \begin{pmatrix} u & \rho & 0 \\ p_\rho/\rho & u & p_S/\rho \\ 0 & 0 & u \end{pmatrix} \tag{2.8.40}$$

and calculate the eigenvalues $\lambda$ from $|A - \lambda E| = 0$.

Solution: $\lambda = u,\ u + \sqrt{p_\rho},\ u - \sqrt{p_\rho}$.

## *2.9 The* DARBOUX *solution of plane waves in non-dissipative gases*

The equations describing plane waves in non-viscous gases can be solved by several methods. In order to understand the equivalence principle and the methods to solve wave equations for dissipative fluids (gas or water) it is of advantage to first understand the solutions for non-dissipative fluids. There are several methods available: solutions using the DARBOUX *equation*, a linear partial differential equation of second order, and the RIEMANN *invariants* [1.1], on which graphic-numerical characteristics methods are based. Another method is a *similarity transformation.*

We consider [2.15] the equations

$$\rho_t + u\rho_x + \rho u_x = 0, \tag{2.9.1}$$

$$u_t + uu_x + \frac{1}{\rho} p_x = 0, \tag{2.9.2}$$

$$p = \frac{\rho^n}{n} a^2, \tag{2.9.3}$$

compare (2.5.29), (2.5.44), (2.5.45); $a$ is a positive constant. These three equations for $\rho, u$ and $p$ can be transformed into the DARBOUX equation, if $n \neq 1$. According to (2.5.44), the condition $n \neq 1$ excludes isothermal waves. Solutions of (2.9.1)–(2.9.3) in cylindrical and spherical coordinates are given in [2.7]. Using the setup

$$\sigma = \frac{a^2 \rho^{n-1}}{n-1} = \frac{n}{n-1} \frac{p}{\rho} \tag{2.9.4}$$

and multiplying (2.9.1) by $a^2 \rho^{n-2}$ one can derive

$$\sigma_t + u\sigma_x + (n-1)\sigma u_x = 0, \tag{2.9.5}$$

$$u_t + uu_x + \sigma_x = 0. \tag{2.9.6}$$

Here $\sigma$ and $u$ depend on $x$ and $t$. We assume that functions $x(u,\sigma)$ and $t(u,\sigma)$ exist. We thus write

$$\mathrm{d}u = u_x \mathrm{d}x + u_t \mathrm{d}t \tag{2.9.7}$$

and

$$\mathrm{d}u = u_x \left(x_u \mathrm{d}u + x_\sigma \mathrm{d}\sigma\right) + u_t \left(t_u \mathrm{d}u + t_\sigma \mathrm{d}\sigma\right). \tag{2.9.8}$$

This gives the equations

$$u_x x_u + u_t t_u = 1, \tag{2.9.9}$$

$$u_x x_\sigma + u_t t_\sigma = 0. \tag{2.9.10}$$

Since

$$\mathrm{d}\sigma = \sigma_x \mathrm{d}x + \sigma_t \mathrm{d}t, \tag{2.9.11}$$

one obtains

$$\sigma_x x_u + \sigma_t t_u = 0, \tag{2.9.12}$$

$$\sigma_x x_\sigma + \sigma_t t_\sigma = 1. \tag{2.9.13}$$

Assuming that $u$ and $\sigma$ are independent of each other, so that the determinant

$$\Delta = \begin{vmatrix} x_u & t_u \\ x_\sigma & t_\sigma \end{vmatrix} \neq 0 \tag{2.9.14}$$

does not vanish, we can solve

$$u_x = t_\sigma/\Delta, \quad u_t = -x_\sigma/\Delta, \quad \sigma_x = -t_u/\Delta, \quad \sigma_t = x_u/\Delta. \tag{2.9.15}$$

Then (2.9.5) and (2.9.6) become

$$x_u - ut_u + (n-1)\sigma t_\sigma = 0, \tag{2.9.16}$$

$$x_\sigma - ut_\sigma + t_u = 0. \tag{2.9.17}$$

This is a linear system and admits *superposition*. We now introduce a new function $V(u,\sigma)$ defined by

$$V_u = x - ut, \quad V_\sigma = -t. \tag{2.9.18}$$

One sees immediately that (2.9.17) is identical with $V_{u\sigma} = V_{\sigma u}$. Thus the partial differential equation for $V$ is integrable. The differential equation for $V$ now reads

$$V_{uu} = V_\sigma + (n-1)\sigma V_{\sigma\sigma}. \qquad (2.9.19)$$

Using new variables $\xi$ and $\eta$ and the substitution

$$\sigma = \xi^2/4, \quad \rightarrow \xi = 2\sqrt{\sigma}, \quad u = \frac{\eta}{\sqrt{n-1}}, \qquad (2.9.20)$$

one obtains the DARBOUX *equation* in the form

$$V_{\xi\xi} + \frac{k}{\xi} V_\xi = V_{\eta\eta}, \quad k = \frac{3-n}{n-1}. \qquad (2.9.21)$$

If $V(\xi, \eta)$ is known, then the solution of (2.9.5) and (2.9.6) is given by

$$\frac{x}{\sqrt{n-1}} - \frac{\eta t}{n-1} = V_\eta, \quad -\frac{\xi t}{2} = V_\xi, \qquad (2.9.22)$$

where (2.9.18) has been used. From (2.9.20) and (2.9.4) one obtains

$$\xi = \frac{2a\rho^{(n-1/2)}}{\sqrt{n-1}}, \quad \eta = u\sqrt{n-1}. \qquad (2.9.23)$$

According to (2.8.3) the adiabatic sonic speed is given by

$$c = \sqrt{\frac{\kappa p}{\rho}} = \frac{\sqrt{\kappa - 1}}{2}\xi = a\rho^{(\kappa-1)/2}. \qquad (2.9.24)$$

Then $\xi \pm \eta$ becomes

$$c \pm \frac{\kappa - 1}{2} u = \text{const.} \qquad (2.9.25)$$

Now we start considerations concerning the solution of the DARBOUX equation. It can be shown that a general solution of it can be found, if the polytropic coefficient $n$ is given by [2.15]

$$n = 1 + \frac{2}{2m+1}, \quad m = 0, 1, 2, \ldots, \qquad (2.9.26)$$

where $m$ is an integer. For adiabatic changes one has

$$\kappa = n \quad \text{or} \quad \kappa = \frac{2m+3}{2m+1}, \quad m = \frac{1}{2} \cdot \frac{3-\kappa}{\kappa - 1}. \qquad (2.9.27)$$

It is well known from statistical physics that a gas with $f$ degrees of freedom has an adiabatic exponent $\kappa = (f+2)/f$. This gives Table 2.2.

Table 2.2. Adiabatic exponents

| fluid | f | $\kappa$ | = | n | m | k |
|---|---|---|---|---|---|---|
| model liquid | – | – | = | −1 | 1 | −2 |
| model gas | 1 | 3 | = | 3.0 | 0 | 0 |
| one-atomic gas | 3 | 5/3 | = | 1.667 | 1 | 2 |
| two-atomic gas | 5 | 7/5 | = | 1.400 | 2 | 4 |
| air | ∼ 5 | – | | 1.405 | ∼ 2 | ∼ 4 |
| water equivalence | – | 2 | = | 2.00 | 0.5 | 1 |

Since the *equivalence theorem* for the solutions of gas waves and water waves states that *tsunamis* have $n = 2$, $m = 0.5$, then solutions of the DARBOUX equations can NOT give solutions of the tsunami wave equation. Thus the paper [2.15] will not help to solve the tsunami wave problem. Solutions of the DARBOUX equation may be found in [2.14], [2.15].

For gases, the RIEMANN *invariants* (RIEMANN *transformation*) [1.1], [2.16] lead to the solutions of the DARBOUX equation, but for large amplitude water waves they produce new solutions. One of these will be discussed in section 2.10. Similarity transformations will be used in section 3.4 (solitons) and later. In [2.18] they again reproduce results offered by the DARBOUX equation.

Although the results obtained in this section seem to be of no interest for the tsunami problem, we will discuss some of the results, because they open some understanding of the effect of rarefaction waves. This type of waves occurs also in the tsunami problem. Conclusions about the development in time are possible without the knowledge of solutions of the DARBOUX equation.

We first write the equations (2.9.5) and (2.9.6) in another form using (2.9.23):

$$\xi_t + \frac{\eta\xi_x}{\sqrt{n-1}} + \frac{\sqrt{n-1}}{2}\xi\eta_x = 0, \quad \eta_t + \frac{\eta\eta_x}{\sqrt{n-1}} + \frac{\sqrt{n-1}}{2}\xi\xi_x = 0. \quad (2.9.28)$$

To be able to discuss properties of the solutions of (2.9.1), (2.9.2) and (2.9.3) without having a solution of (2.9.21), we now rewrite (2.9.28) in the form

$$\frac{\mathrm{d}\xi}{\mathrm{d}t} = \xi_x\left(\frac{\mathrm{d}x}{\mathrm{d}t} - \frac{\eta}{\sqrt{n-1}}\right) - \frac{\sqrt{n-1}}{2}\xi\eta_x,$$

$$\frac{\mathrm{d}\eta}{\mathrm{d}t} = \eta_x \left( \frac{\mathrm{d}x}{\mathrm{d}t} - \frac{\eta}{\sqrt{n-1}} \right) - \frac{\sqrt{n-1}}{2} \xi \xi_x. \tag{2.9.29}$$

This system can again be transformed using two new variables $v$ and $w$ (of the type of RIEMANN *invariants*, see [1.1])

$$v = \xi + \eta, \quad w = \xi - \eta. \tag{2.9.30}$$

The result is

$$\begin{aligned} \frac{\mathrm{d}v}{\mathrm{d}t} &= v_x \left( \frac{\mathrm{d}x}{\mathrm{d}t} - \frac{\eta}{\sqrt{n-1}} - \frac{\sqrt{n-1}}{2} \xi \right), \\ \frac{\mathrm{d}w}{\mathrm{d}t} &= w_x \left( \frac{\mathrm{d}x}{\mathrm{d}t} - \frac{\eta}{\sqrt{n-1}} + \frac{\sqrt{n-1}}{2} \xi \right). \end{aligned} \tag{2.9.31}$$

To find an extremum, we assume $\mathrm{d}v/\mathrm{d}t = 0$ and $\mathrm{d}w/\mathrm{d}t = 0$. We then obtain from (2.9.31)

$$\frac{\mathrm{d}x}{\mathrm{d}t} = \frac{2\eta \pm (n-1)\xi}{2\sqrt{n-1}} = u \pm a\rho^{(n-1)/2}. \tag{2.9.32}$$

Now we consider a special constant value $A$ of $v$ and $B$ of $w$. Then according to (2.9.32) the value $v = A$ propagates with the wave speed to the right

$$\frac{\mathrm{d}x}{\mathrm{d}t} = \frac{2A + (n-3)\xi}{2\sqrt{n-1}} \tag{2.9.33}$$

and the value $w = B$ propagates with

$$\frac{\mathrm{d}x}{\mathrm{d}t} = -\frac{2B + (n-3)\xi}{2\sqrt{n-1}} \tag{2.9.34}$$

to the left. We remark that both propagation speeds are constant for $n = 3$. For adiabatic behavior $n = \kappa$ one obtains with (2.9.24)

$$\frac{\mathrm{d}x}{\mathrm{d}t} = u + c, \quad \text{for } v = A = \text{const}, \tag{2.9.35}$$

$$\frac{\mathrm{d}x}{\mathrm{d}t} = u - c, \quad \text{for } w = B = \text{const}. \tag{2.9.36}$$

Here $u$ and $c$ are not constants, but depend on $x$ and $t$. Of great interest is now the fact that the waves steepen up, that means that greater $v$-values (greater densities) propagate faster than smaller $v$-values: a *shock wave* develops. On the other hand waves producing a density decrease (*rarefaction wave*) become more smooth – no *rarefaction shock wave* is possible.

A shock is a discontinuity described by equations (2.8.31)–(2.8.33). At such discontinuities the differential equations no longer have meaning. In reality the shock is however not a discontinuity at all but a narrow zone, a few mean free paths in thickness through which the variables change very steeply but continuously. In the theory presented here up to now heat conduction and viscosity effects have been neglected. They produce an entropy increase across the shock ($S_2 > S_1$). Then (2.9.1)–(2.9.3) has to be replaced by the system (2.8.37)

$$\rho_t + \rho_x u + u_x \rho = 0, \tag{2.9.37}$$

$$\rho_t u + u \rho_x + \frac{\partial}{\partial x}\left(\rho u^2 + p - \frac{4}{3}\eta u_x\right) = 0, \tag{2.9.38}$$

$$\frac{\partial}{\partial t}\left(\rho\left(U + \frac{1}{2}u^2\right)\right) + \frac{\partial}{\partial x}\left(\rho u\left(U + \frac{1}{2}u^2\right) + u\left(p - \frac{4}{3}\eta u_x\right) - \lambda T_x\right) = 0, \tag{2.9.39}$$

or

$$\rho T\left(S_t + u S_x\right) - \frac{4}{3}\eta u_x^2 - \frac{\partial}{\partial x}\left(\lambda T_x\right) = 0, \tag{2.9.40}$$

compare another formulation (2.5.35)–(2.5.39) and (2.5.48).

In order to be able to treat shocks by numerical methods, an *artificial viscosity* has been introduced (LAX *viscosity*) [2.20] [2.13].

We now give solutions to the DARBOUX equation. Thus we have to assume that (2.9.26) is valid [2.15]. For $m = 0$ one has $k = 0, n = 3$ (Table 2.2) and (2.9.21) has the solution

$$V = f(\xi + \eta) + g(\xi - \eta), \tag{2.9.41}$$

where $f$ and $g$ are arbitrary functions. They are determined by the initial conditions for $\rho$ and $u$. (2.9.41) is a solution of the linear D'ALEMBERT *equation* of acoustics. We see that the values $k = 0, n = 3$ initiate a quasi-linearization.

For $k = 2m$ various solutions have been given in [2.15]. Some of them are identical with the RIEMANN solutions [2.16]. For general $m$ solutions may be found in [2.13].

## Problems

1. Determine the type of the DARBOUX equation.

   Solution: use (1.2.1), (1.2.7). The equation is parabolic.

2. Is it possible to solve (2.9.21) by using $V(\xi, \eta) = X(\xi) \cdot E(\eta)$?

3. Is it possible to solve (2.9.21) by a similarity setup $V(\xi, \eta) = \xi^\alpha \eta^\beta$ (BOLTZMANN *transformation*)?

4. Instead of the pressure $p(x,t)$, given by (2.9.3), RIEMANN in [2.16] uses a function $\varphi(\rho) = p(\rho)$, so that

$$\varphi'(\rho)\frac{\partial \rho}{\partial x} = \frac{\partial p}{\partial x}. \tag{2.9.42}$$

   Write (2.9.1) and (2.9.2) in a new form using (2.9.42).

   Solution:

$$\frac{\partial u}{\partial t} + u\frac{\partial u}{\partial x} = -\varphi'(\rho)\frac{\partial \log \rho}{\partial x}, \tag{2.9.43}$$

$$\frac{\partial \log \rho}{\partial t} + u\frac{\partial \log \rho}{\partial x} = -\frac{\partial u}{\partial x}. \tag{2.9.44}$$

5. Multiply (2.9.44) by $\pm\sqrt{\varphi'(\rho)}$ and add it to (2.9.43), then

$$f(\rho) = \int \sqrt{\varphi'(\rho)}\mathrm{d}\log \rho, \tag{2.9.45}$$

$$f(\rho) + u = 2r, \quad f(\rho) - u = 2s \tag{2.9.46}$$

   transforms the equations. Hint: calculate $\partial r/\partial t$ and $\partial s/\partial t$.

   Solution [2.16]:

$$\mathrm{d}r = \frac{\partial r}{\partial x}\left(\mathrm{d}x - (u + \sqrt{\varphi'(\rho)}\,)\mathrm{d}t\right), \tag{2.9.47}$$

$$\mathrm{d}s = \frac{\partial s}{\partial x}\left(\mathrm{d}x - (u - \sqrt{\varphi'(\rho)}\,)\mathrm{d}t\right). \tag{2.9.48}$$

   The quantities $r, s$ are often called RIEMANN *invariants*.

6. RIEMANN transforms the equations (2.9.47) and (2.9.48) into the form

$$\mathrm{d}r = \frac{\partial r}{\partial x}\left\{\mathrm{d}\left(x - \left(u + \sqrt{\varphi'(\rho)}\right)t\right)\right.$$
$$\left. + \left[\mathrm{d}r\left(\frac{\mathrm{d}\log\sqrt{\varphi'(\rho)}}{\mathrm{d}\log\rho} + 1\right) + \mathrm{d}s\left(\frac{\mathrm{d}\log\sqrt{\varphi'(\rho)}}{\mathrm{d}\log\rho} - 1\right)\right]t\right\}, \quad (2.9.49)$$

$$\mathrm{d}s = \frac{\partial s}{\partial x}\left\{\mathrm{d}\left(x - \left(u - \sqrt{\varphi'(\rho)}\right)t\right)\right.$$
$$\left. - \left[\mathrm{d}s\left(\frac{\mathrm{d}\log\sqrt{\varphi'(\rho)}}{\mathrm{d}\log\rho} + 1\right) + \mathrm{d}r\left(\frac{\mathrm{d}\log\sqrt{\varphi'(\rho)}}{\mathrm{d}\log\rho} - 1\right)\right]t\right\}. \quad (2.9.50)$$

Assume that $r$ and $s$ are independent variables and derive linear differential equations for $x$ and $t$.

Solution:

$$\frac{\partial\left(x - \left(u + \sqrt{\varphi'(\rho)}\right)t\right)}{\partial s} = -t\left(\frac{\mathrm{d}\log\sqrt{\varphi'(\rho)}}{\mathrm{d}\log\rho} - 1\right), \quad (2.9.51)$$

$$\frac{\partial\left(x - \left(u - \sqrt{\varphi'(\rho)}\right)t\right)}{\partial r} = t\left(\frac{\mathrm{d}\log\sqrt{\varphi'(\rho)}}{\mathrm{d}\log\rho} - 1\right) \quad (2.9.52)$$

and receive the total differential

$$\left(x - \left(u + \sqrt{\varphi'(\rho)}\right)t\right)\mathrm{d}r - \left(x - \left(u - \sqrt{\varphi'(\rho)}\right)t\right)\mathrm{d}s. \quad (2.9.53)$$

The solution of this differential now satisfies the partial differential equation for $w(r,s)$

$$\frac{\partial^2 w}{\partial r \partial s} = m\left(\frac{\partial w}{\partial r} + \frac{\partial w}{\partial s}\right) = -t\left(\frac{\mathrm{d}\log\sqrt{\varphi'(\rho)}}{\mathrm{d}\log\rho} - 1\right), \quad (2.9.54)$$

where

$$m = \frac{1}{2\sqrt{\varphi'(\rho)}}\left(\frac{\mathrm{d}\log\sqrt{\varphi'(\rho)}}{\mathrm{d}\log\rho} - 1\right). \quad (2.9.55)$$

7. Assume adiabatic behavior $\varphi(\rho) = a\rho^\kappa$ and calculate $\sqrt{\varphi'(\rho)}$ and $m$.

   Solution:

$$\sqrt{\varphi'(\rho)} + u = \frac{\kappa+1}{2} r + \frac{\kappa-3}{2} s, \tag{2.9.56}$$

$$\sqrt{\varphi'(\rho)} - u = \frac{\kappa-3}{2} r + \frac{\kappa+1}{2} s, \tag{2.9.57}$$

$$m = \frac{\kappa-3}{2(\kappa-1)(r+s)} = \left(\frac{1}{2} - \frac{1}{(\kappa-1)}\right)\frac{1}{\sigma}, \tag{2.9.58}$$

   where

$$\sigma = r + s = \frac{2a\sqrt{\kappa}}{\kappa-1}\rho^{(\kappa-1)/2}. \tag{2.9.59}$$

8. Using

$$\lambda = \frac{1}{2} - \frac{1}{\kappa-1}, \tag{2.9.60}$$

   gives a solution of (2.9.54) in the form of a *hypergeometric function* [1.1]

$$y(z) = F(1+\lambda, -\lambda, 1, z). \tag{2.9.61}$$

   Plot this function using Mathematica.

## *2.10 The equivalence theorem*

This theorem states that solutions for waves in gases can be reinterpreted as solutions for large water waves. To prove this for a compressible flow we establish the basic equations for water. We first consider a steady two-dimensional flow bounded by two vertical walls. For a water prism with a base $\mathrm{d}x\mathrm{d}y$ and a varying height $0 \leq h \leq h_0$ in the $z$-direction, the water mass contained within the prism is given by $\mathrm{d}x\mathrm{d}yh\rho_0$, where $\rho_0$ is the (constant) water density. Here $h(x, y)$ designates the location of a water layer in the $z$-direction. $z = 0$ designates the water surface and $h_0$ describes the distance between the unperturbed surface and the plane buttom of the channel. If we designate by $u$ and $v$ the outflow velocity in the $x$- and $y$-direction respectively, then the total inflow $\mathrm{d}q_e$ must be equal to the total outflow $\mathrm{d}q_a$ to conserve the mass in the prism. The total inflow in two directions is given by

$$\mathrm{d}q_e = uh\mathrm{d}y + vh\mathrm{d}x. \tag{2.10.1}$$

Since $u$ and $h$ depend on space, the outflow at another location is

$$\mathrm{d}q_a = \left(u + \frac{\partial u}{\partial x}\mathrm{d}x\right)\left(h + \frac{\partial h}{\partial x}\mathrm{d}x\right)\mathrm{d}y + \left(v + \frac{\partial v}{\partial y}\mathrm{d}y\right)\left(h + \frac{\partial h}{\partial y}\mathrm{d}y\right)\mathrm{d}x. \tag{2.10.2}$$

Neglecting higher terms like $(\mathrm{d}x)^2$ etc, one obtains the balance $\mathrm{d}q_a - \mathrm{d}q_e = 0$ in the form of a *continuity equation* [2.21]

$$\frac{\partial(hu)}{\partial x} + \frac{\partial(hv)}{\partial y} = 0. \tag{2.10.3}$$

Compare with the gasdynamic equation (2.2.4)

$$\frac{\partial(\rho u)}{\partial x} + \frac{\partial(\rho v)}{\partial y} = 0. \tag{2.10.4}$$

One sees immediately that the water continuity equation has exactly the same form as the continuity equation for a fictive gas with density $\hat{\rho} = \rho_0 h$, where $\rho_0$ is the (constant) water density. Inserting $h = \hat{\rho}/\rho_0$ into (2.10.3) yields

$$\frac{\partial(\hat{\rho} u)}{\partial x} + \frac{\partial(\hat{\rho} v)}{\partial y} = 0. \tag{2.10.5}$$

Now we consider the BERNOULLI *equation* (2.3.20) which reads

$$\frac{\rho_0 w^2}{2} + p + \rho_0 g z = \text{ const } = \frac{\rho_0 w_1^2}{2} + p_1 + \rho_0 g z_1, \tag{2.10.6}$$

where $w^2 = u^2 + v^2$. It expresses that the sum of potential plus kinetic energy is constant. The values $w_1, p_1$ are arbitrary values valid at $z = z_1$. Now we specify this point. For $z = z_0$ the water depth $h$ is equal to the largest depth $h_0$, where $w = w_0 = 0$. We thus replace $p_1 \to p_0$, $z_1 \to z_0$, $w_1 \to w_0 = 0$. Then (2.10.6) yields

$$\frac{\rho_0 w^2}{2} + p + \rho_0 g z = p_0 + \rho_0 g z_0. \tag{2.10.7}$$

This yields

$$w^2 = 2g(z_0 - z) + 2(p_0 - p)/\rho_0. \tag{2.10.8}$$

Now the pressure $p$ in streaming water may be assumed to be a linear function of the distance between the actual position $z$ and the free water surface:

$$p = \rho_0 g(h - z), \quad p_0 = \rho_0 g(h_o - z_0). \tag{2.10.9}$$

Inserting these expressions into (2.10.8) yields

$$w^2 = 2g(h_0 - h). \tag{2.10.10}$$

Since the maximum height difference $h_0 - h$ is given by $h_0$, one has

$$w_{\max} = \sqrt{2gh_0}, \tag{2.10.11}$$

compare with (2.3.21)! Let us compare (2.10.11) with $v_{\max}$ for gases. According to (2.5.51) one has

$$v_{\max} = \sqrt{\frac{2\kappa p_0}{(\kappa - 1)\rho_0}}. \tag{2.10.12}$$

Here $p_0$ and $\rho_0$ are pressure and density of an ideal gas at rest. Now we specialize the equation of motion (2.3.13) for a gas which is free from exterior forces ($\vec{g} = \vec{g}\,' = 0$), which is conservative-adiabatic ($\nu = 0$). The flow considered should be steady ($\partial/\partial t = 0$) and one-dimensional. Then (2.3.13) reads for the fictive gas streaming in the $x$-direction

$$v\frac{\partial v}{\partial x} = -\frac{1}{\hat{\rho}}\frac{\partial \hat{p}}{\partial x}. \tag{2.10.13}$$

Defining a pressure

$$\hat{p} = \int_0^h p\mathrm{d}z = \frac{\rho_0 g h^2}{2} = \frac{g\hat{\rho}^2}{2\rho_0} \tag{2.10.14}$$

for the fictive gas $\hat{\rho} = \rho_0 h$, we obtain from (2.10.13) the equation of motion for water ($\rho_0$)

$$v\frac{\partial v}{\partial x} = -g\frac{\partial h}{\partial x}. \tag{2.10.15}$$

The fictive gas $\hat{\rho}, \hat{p}$ described by (2.10.5) and (2.10.13) is adiabatic. Hence the adiabatic equation (2.5.42) is valid and one has

$$p = \frac{p_0}{\rho_0^\kappa}\rho^\kappa = \mathrm{const}\cdot\rho^\kappa. \tag{2.10.16}$$

For the fictive gas representing water, (2.10.14) yields the "equation of state"

$$\hat{p} = \frac{g}{2\rho_0}\hat{\rho}^2 = \mathrm{const}\cdot\hat{\rho}^2 = \frac{\rho_0 g h^2}{2}. \tag{2.10.17}$$

Thus we find that the fictive gas representing water has

$$\kappa = 2. \tag{2.10.18}$$

We derived this result for a steady two-dimensional flow. The same result can also be obtained for an unsteady (time dependent) one-dimensional flow [2.10], see for instance problem 4.

Thus the *equivalence theorem* may be formulated as follows:

1. Write down the gasdynamic equations (2.5.34), (2.5.35) and (2.5.38) and replace $\rho \to \hat{\rho}$, $p \to \hat{p}$ to obtain modified equations like (2.10.41) etc.

2. Use $\hat{\rho} = \rho_0 h$, $\hat{p} = \rho_0 g h^2/2$ from (2.10.14) and insert into the modified equations, see problem 5 in this section.

For a streaming compressible fluid subsonic flow ($c < v$) and supersonic flow ($c > v$) have been discussed in section 2.8 and *shock waves* appeared. For a fictive gas the same phenomena have to be expected. We may rewrite the shock equations (2.8.25), (2.8.26) in the form

$$h_1 u_1 = h_2 u_2, \tag{2.10.19}$$

$$\frac{g h_1^2}{2} + u_1^2 h_1 = \frac{g h_2^2}{2} + u_2^2 h_2. \tag{2.10.20}$$

Expressing $u_1$ and $u_2$ by $h_1$ and $h_2$ we have

$$u_1^2 = \frac{g h_2 (h_1 + h_2)}{2 h_1}, \quad u_2^2 = \frac{g h_1 (h_1 + h_2)}{2 h_2}. \tag{2.10.21}$$

Since $u_1^2 \neq u_2^2$ the kinetic energy is not conserved over the shock. The difference has been dissipated in the shock. If the fluid flows from the location 1 with lower height $h_1$ to the location of greater height, if

$$h_2 > h_1, \tag{2.10.22}$$

then (2.10.21) yields

$$u_1 > \sqrt{g h_1}, \quad u_2 < \sqrt{g h_2}. \tag{2.10.23}$$

The *subsonic (subcritical) flow* $u_2 < \sqrt{g h_2}$ corresponds to water streaming and the *supersonic (supercritical)* flow $u_1 > \sqrt{g h_1}$ to *shooting flow* (supercritical flow). For the shock the HUGONOT state equation (2.8.30) becomes

$$\left(\frac{h_2}{h_1}\right)^3 - 3\left(\frac{h_2}{h_1}\right)^2 + 3\frac{h_2}{h_1} - 1 = 0. \tag{2.10.24}$$

As will be shown in problem 2, this indicates that (2.8.30) is not valid for water jump.

Now we investigate the equivalence theorem for one-dimensional unsteady dissipative flow. The equations (2.9.37) - (2.9.40) describe such a gas flow. (2.9.40) describes entropy increase, so that the flow is no longer polytropic. The initial state $\rho_1, u_1, T_1, S_1$ and the final state $\rho_2, u_2, T_2, S_2$ are connected by an increase $S_2 - S_1 > 0$ of entropy. State 1 goes over in state 2 by an irreversible process. But it is always possible to setup a polytropic replacement process described by $p_1 V_1^n = p_2 V_2^n$. We therefore assume that $n = \kappa = 2$ is still valid for a dissipative process too.

We now solve a one-dimensional unsteady flow of water using directly the hydrodynamic equations but the gasdynamic characteristics method. We investigate the sudden rapture of an embankment dam.

At the occasion of a rupture, a water surge shall propagate into the channel downstream of the dam. Such surge may generate heavy destructions along the channel or river. The bottom of the channel is a horizontal plane. Let us assume that the channel has a width $B$ and extends in the direction of the $x$-axis. Let the water depth $H$ in the storage lake be $H = 2.2$ m and the water depth $h$ in the channel is assumed to be $h_0 = 1.2$ m. The water level after a dam rupture will be designated by $h(x,t)$. Thus the local water mass is given by $\rho_0 B h(x,t) = q(x,t)\rho_0$, where $\rho_0$ is the (constant) water density. Let the storage lake have an extension of 5000 m in the $x$-direction. The dam itself may be situated at $x = 5000$ m and rupture may occur suddenly at $t = 0$. Then we have the initial conditions for $h$ and the stream velocity $u(x,t)$ at $t = 0$:

$$
\begin{aligned}
h(x,0) &= H = 2.2 \text{ m}, \quad u(x,0) = 0, \quad 0 \leq x \leq 5000 \text{ m},\\
h(x,0) &= h_0 = 1.2 \text{ m}, \quad u(x,0) = 0, \quad 5\,000 \leq x \leq \infty. \qquad (2.10.25)
\end{aligned}
$$

This indicates that at $t = 0$ and $x = 5000$ m a vertical water wall of a height $H - h_0 = 1$ m exists. At the other end of the lake ($x = 0$) no flow is present. The relevant equations describing the evolution in time of these nonlinear one-dimensional phenomena are:

the continuity equation

$$
\frac{\partial}{\partial t}(\rho_0 q(x,t)) + \frac{\partial}{\partial x}(\rho_0 u(x,t) q(x,t)) = 0, \quad q_t + u_x q + q_x u = 0, \qquad (2.10.26)
$$

and the equation of motion

$$
\rho_0 u_t + \rho_0 u u_x + p_x = 0. \qquad (2.10.27)
$$

The local hydrostatic pressure $p(x,t)$ per unit length is given by

$$
p(x,t) = \rho_0 g q(x,t)/B. \qquad (2.10.28)
$$

Then we can now write for (6.4.3)

$$u_t + uu_x + \frac{g}{B}q_x = 0. \tag{2.10.29}$$

We thus have two quasilinear partial differential equations (2.10.26) and (2.10.29) for the two unknown functions $u(x,t)$ and $q(x,t)$. We use the method of characteristics developed in section 1.4 for such a system of two partial differential equations of first order. We compare our system of two partial equations with (1.4.18) and read off

$$\begin{aligned} &a_{11} = 1, \quad a_{12} = 0, \quad a_{21} = 0, \quad a_{22} = 1, \\ &b_{11} = u, \quad b_{12} = g/B, \, b_{21} = q, \quad b_{22} = u. \end{aligned} \tag{2.10.30}$$

Here we have used the following substitutions in (1.4.18) $x \to t$, $y \to x$, $v \to q$ to obtain (2.10.26) and (2.10.29). Then (1.4.23) yields the propagation speed of *small* amplitude waves

$$\begin{aligned} \frac{\mathrm{d}x}{\mathrm{d}t} &= u + \sqrt{\frac{gq}{B}} = u + \sqrt{gh} \quad \text{(downstream)}, \\ \frac{\mathrm{d}x}{\mathrm{d}t} &= u - \sqrt{\frac{gq}{B}} = u - \sqrt{gh} \quad \text{(upstream)} \end{aligned} \tag{2.10.31}$$

and equations (1.4.24) and (1.4.25) result in

$$\pm\sqrt{\frac{gq}{B}}\mathrm{d}u + \frac{g}{B}\mathrm{d}q = 0. \tag{2.10.32}$$

To obtain this result observe the substitutions and use equation (1.4.25) for $V_2$. Insert for $k' = \mathrm{d}x/\mathrm{d}t$ from (2.10.31). The two equations describe the modification of the *state variables* $u, q$ along the characteristics (2.10.31). The problem is now that we cannot use or integrate the characteristics equation because they contain the still unknown solutions $u(x,t)$ and $q(x,t)$. We first make a transformation to a new variable $\lambda$ [ms$^{-1}$]. We define

$$\mathrm{d}\lambda = \sqrt{\frac{g}{B}}\frac{\mathrm{d}q}{\sqrt{q}}, \quad \lambda(q) = \int_0^q \frac{\mathrm{d}q}{\sqrt{q}}\sqrt{\frac{g}{B}} = \int_0^h \sqrt{\frac{g}{h}}\,\mathrm{d}h = 2\sqrt{gh}. \tag{2.10.33}$$

Then we use the RIEMANN *invariants* defined by

$$r = u + \lambda \quad s = u - \lambda, \quad u = (r+s)/2, \quad \lambda = (r-s)/2, \tag{2.10.34}$$

$$\mathrm{d}u \pm \mathrm{d}\lambda = 0, \quad u \pm \lambda = const = \begin{cases} r \\ s \end{cases}. \tag{2.10.35}$$

The $r, s$ or $u, q$ plane is called *state plane* by some authors. We now consider a mapping between the "linear" state plane $r, s$ and the nonlinear physical plane described by $x, t$. Let us discuss the correspondence between the two planes. We allocate the point $P(r_1, s_1)$ of the state plane to the dam location point $\bar{P}$ (5000,0) of the physical plane. This expresses the fact that in the point $\bar{P}(x = 5000, t = 0)$ a local water wall of absolute height $h(5000, 0) = H = h_1$ or relative height 1 m above the normal water level in the channel exists with streaming velocity $u(5000, 0) = u_1 = 0$. According to (2.10.33) the height $h_1 = H = 2.2$ m corresponds to $r_1 = u_1 + \lambda_1 = \lambda_1$, $s_1 = -\lambda_1$, $\lambda_1 = 2\sqrt{gh_1}$. At the other end of the lake $x = 0$, one has $u(0,0) = 0$ and $h(0,0) = H = 2.2$ m. The point $\bar{Q}(0,0)$ corresponds to $Q(r_Q s_Q)$, where $u_Q = 0, \lambda_Q = 2\sqrt{gH}$, $r_Q = \lambda_Q$, $s_Q = -\lambda_Q$, $r_Q = -s_Q$. Inserting numbers for $h, H$ and $g$ (9.81 ms$^{-2}$) we receive for the point $\bar{P}$

$$\lambda_1 = 2\sqrt{2.2 \cdot 9.81} = 9.291, \;\; u_1 = 0, \;\; r_1 = 9.291, \;\; s_1 = -9.291, \qquad (2.10.36)$$

all measured in [ms$^{-1}$]. On the other end of the lake we have for $\bar{Q}(0,0)$

$$\lambda_Q = 9.291, \; u_Q = 0, \;\; r_Q = 9.291, \; s_Q = -9.291. \qquad (2.10.37)$$

At the time $t = 0$ of the rupture of the dam the same physical states exist at $x = 5000$ and $x = 0$. But, at this time, the dam breaks down and elementary waves (composing later on a steepening surge downstream and a rarefaction wave upstream) start at $x = 5000$. Replacing in (2.10.31) the $\mathrm{d}x \to \Delta x$, $\mathrm{d}t \to \Delta t$ we can write for the wave speeds

$$\frac{\Delta x}{\Delta t} = u \pm \sqrt{gh} = u \pm \frac{\lambda}{2}. \qquad (2.10.38)$$

Thus, the first elementary wave running to the left to $x = 0$ and upstream reduces the water level $H$ in the lake. It has a wave speed $\Delta x/\Delta t = 0 - \lambda_1/2 = -4.646$[ms$^{-1}$] and $s_1 = -9.291 = const$, $r_1 = +9.291$. The wave running to the right (downstream) increases the water level $h(x,t)$ in the channel from $h_0 = h_1$ to $h_2$ and has a wave speed $\Delta x/\Delta t = 0 + \lambda_1/2 = +4.646$[ms$^{-1}$] and $s_1 = -9.291$, $r_1 = +9.291 = const$. Waves running to the left from $P$ to $Q$ transfer their $s$-value to $Q$, since $s = const$ is valid for waves running to the left: $s_Q = s_1$. Waves running to the right, downstream from $P$ to $x \to \infty$ transfer their $r$-value, so that the whole domain right-hand of the dam ($x > 5000$) always has the same $r$-value. At the time $t = 0 + \Delta t$ the next two elementary waves start. Both waves now run into domains where the states had been modified by the first two waves. The upstream wave enters an area where the water level had been reduced from $H$ to $H - h_1$ and might be reflected at the lake end $x = 0$. It will no longer return with 4.646 [ms$^{-1}$], because water level and driving pressure had been lowered. The second elementary downstream wave starting at $t = \Delta t$ will be faster than the first one because in the channel the water level had been increased by the first downstream wave and the water had started to stream to $x \to \infty$. In order to be able to calculate the $\Delta u, \Delta\lambda$, etc., we need to have some knowledge about the final state, $t \to \infty, x \to \infty$. For an infinitely long channel we define the point $R(r_N, s_N)$ in the state plane. Apparently the final conditions will read for $R$: $q(x, t \to \infty) = h_0 = 1.2\text{m}$, $\lambda_N = 2 \cdot \sqrt{1.2 \cdot 9.81} = 6.862$, and for $\bar{R}(\infty, \infty)$ $u_N = 0$, $r_N = 6.862$, $s_N = -6.862$. The whole phenomenon of the break-down of the dam occupies a square in the $s, r$ plane. The four corners are given by $R(s_N = -6.862, r_N = +6.862)$, $(s_N = -6.862, r_1 = 9.291)$, $(s_1 = -9.291, r_N = 6.862)$ and $P = Q(s_1 = -9.291, r_1 = 9.291)$. Now it is the accuracy and our will that have to decide how many steps $i = 1 \ldots N$ we will calculate. For this decision we consider the pressure difference from $p_1 = \rho_0 g q(x, 0) = \rho_0 g B h(x, 0) = \rho_0 B \lambda_1^2/4 = \rho_0 B r_1^2/4$ down to $p_N = B r_N^2/4$. This concerns the variation $\lambda_1 \to \lambda_N$, $r_1 \to r_N$ etc. If we choose $N = 10$ pressure steps, then each elementary wave carries $\Delta r = (9.921 - 6.862)/10 = 0.243 = |\Delta s|$. This corresponds to an accuracy of 2.6 % (0.243:9.291). Table 2.3 describes the situation in detail.

Table 2.3. Pressure steps (for a downstream wave)

| | in front of the wave | | | | behind the wave | | |
|---|---|---|---|---|---|---|---|
| Nr | $s$ | $\lambda$ | $u$ | $\Delta x/\Delta t$ | $s$ | $\lambda$ | $u$ |
| 1 | −9.291 | +9.291 | 0 | 4.646 | −9.048 | +9.170 | +0.122 |
| 2 | −9.048 | +9.170 | +0.122 | 4.707 | −8.805 | +9.048 | +0.243 |
| 3 | −8.805 | +9.048 | +0.243 | 4.767 | −8.562 | +8.927 | +0.365 |
| 4 | −8.562 | +8.927 | +0.365 | 4.829 | −8.319 | +8.805 | +0.486 |
| 5 | −8.319 | +8.805 | +0.486 | 4.889 | −8.076 | +8.684 | +0.608 |
| 6 | −8.076 | +8.684 | +0.608 | 4.950 | −7.833 | +8.562 | +0.729 |
| 7 | −7.833 | +8.562 | +0.729 | 5.010 | −7.590 | +8.441 | +0.851 |
| 8 | −7.590 | +8.441 | +0.851 | 5.072 | −7.347 | +8.319 | +0.936 |
| 9 | −7.347 | +8.319 | +0.936 | 5.096 | −7.104 | +8.198 | +1.094 |
| 10 | −7.104 | +8.198 | +1.094 | 5.193 | −6.862 | +8.077 | +1.215 |
| 11 | −6.862 | +8.007 | +1.215 | 5.254 | [−6.619 | +7.955 | +1.336] |

The grid of points within the square in the $r, s$ plane can be mapped into the $x, t$ plane: for every point in the $r, s$ plane the values $u(x,t), q(x,t)$ in the $x, t$ plane are defined by the equations (2.10.31) to (2.10.35). Interpolation within the grid delivers any wanted $u(x,t), q(x,t)$ and thus the solution of equations (2.10.26)–(2.10.29).

This example demonstrates the solution of a nonlinear hyperbolic problem of water flow. Similar methods can be used to solve the tsunami problem.

## Problems

1. Derive (2.10.15) from (2.10.13). Hint: use $\hat{\rho} = \rho_0 \cdot h$, $\rho_o$ (water) = const.

2. Derive (2.10.19).
   Hint: use the formal replacement $p \to \hat{p} = g\hat{\rho}^2/2\rho_0$, $\rho \to \hat{\rho} = \rho_0 h$, $\kappa = 2$ and insert into (2.8.30). Use $h_2/h_1 \to x$. The Mathematica command:

   ```
   Solve[x^3-3*x^2+3*x-1==0,x]
   ```

   yields the result $x_1 = 1 = x_2 = x_3$, $h_1 = h_2$.

3. Write (2.9.40) in the form

   $$T\frac{\mathrm{d}S}{\mathrm{d}t} = \frac{1}{\rho}\left(\frac{4}{3}\eta u_x^2 + \lambda T_{xx}\right) = Q$$

   and integrate.

   Solution: $t = \exp\left((S - S_0)/Q\right)$.

4. Derive the water wave equations for an unsteady one-dimensional flow [2.10].

   Solution:

   water

   $$\frac{\partial h}{\partial t} + \frac{\partial (vh)}{\partial x} = 0 \qquad (2.10.39)$$

   $$\frac{\partial v}{\partial t} + v\frac{\partial v}{\partial x} = -g\frac{\partial h}{\partial x} \qquad (2.10.40)$$

   fictive gas

   $$\frac{\partial \hat{\rho}}{\partial t} + \frac{\partial v\hat{\rho}}{\partial x} = 0 \qquad (2.10.41)$$

   $$\frac{\partial v}{\partial t} + v\frac{\partial v}{\partial x} = -\frac{1}{\hat{\rho}}\frac{\partial \hat{p}}{\partial x}. \qquad (2.10.42)$$

   Here $\hat{\rho} = \rho_0 h$, $\hat{p} = (\rho_0 g h^2)/2$ have been used.

5. Derive the equivalent equations for water under the assumption of dissipation due to viscosity and heat conduction. Neglect exterior forces and obtain the partial differential equations for an unsteady one-dimensional flow in the $x$-direction over a channel with constant depth.

   Solution:

$$\frac{\mathrm{d}h}{\mathrm{d}t} + h\frac{\partial v}{\partial x} = 0, \tag{2.10.43}$$

   compare (2.5.35);

$$h\frac{\mathrm{d}v}{\mathrm{d}t} = -gh\frac{\partial h}{\partial x} + \frac{\eta}{\rho_0}v_{xx}, \tag{2.10.44}$$

   compare (2.5.36);

$$h\frac{\mathrm{d}}{\mathrm{d}t}\left(\alpha T + \frac{v^2}{2}\right) + v_x\frac{g}{2}h^2 + vghh_x = \frac{\eta}{\rho_0}\frac{\partial}{\partial x}(vv_x) + \frac{\lambda}{\rho_0}T_{xx}, \tag{2.10.45}$$

   compare (2.5.28). Here we used

$$\alpha = \frac{C}{\kappa - 1} = C, \quad \kappa = 2.$$

   The thermodynamic equation of state is given by the two definitions for $\hat{\rho}$ and $\hat{p}$ above and $\kappa = 2$.

   Finally we consider a replacement adiabatic ($\kappa = 2$) for the initial and final states describing the irreversible process $1 \longrightarrow 2$

$$\frac{p_1}{p_2} = \left(\frac{\rho_1}{\rho_2}\right)^{\kappa} \rightarrow \frac{\hat{p}_1}{\hat{p}_2} = \left(\frac{\hat{\rho}_1}{\hat{\rho}_2}\right)^{\kappa} \rightarrow \frac{h_1^2}{h_2^2} = \left(\frac{h_1}{h_2}\right)^{\kappa}, \quad \rightarrow \kappa = 2. \tag{2.10.46}$$

   We thus have three equations (2.10.43), (2.10.44) and (2.10.45) for the three variables $h, v, T$ to describe the flow.

6. Derive the characteristics for the two quasilinear equations (2.10.39) and (2.10.40). Use (1.4.18) and read the $a_{ik}, b_{ik}$ from (2.10.39) and (2.10.40).

7. Derive the characteristics for the system of the three quasilinear partial differential equations (2.10.43), (2.10.44) and (2.10.45). Use (1.4.26).

8. In problem 14 of section 2.5 we investigated the pressure-curve $p(v)$ for a compressible gas flow. Using nondimensional quantities $y =$

$v/v_m$, $x = p/p_0$ we derived the pressure-curve $p(v)$, see Fig. 2.4. Replacing again $v/v_m = y$, but now $h/h_0 = x$ one may derive the pressure-curve $h(v)$ for water.

Solution: Use (2.5.55) for $\kappa = 2$ which gives $y = (1 - x^2)^2$. Plot this function, investigate the inflection point and compare with Fig. 2.4 in section 2.5.

9. Derive a *potential equation* $\varphi(x, y)$ from (2.10.10) and (2.10.3).

   Solution: $w^2 = u^2 + v^2$ and (2.10.10) yields

$$h_x = -\frac{1}{g}(uu_x + vv_x), \quad h_y = -\frac{1}{g}(uu_y) + vv_y). \tag{2.10.47}$$

   Insert into (2.10.3) to obtain after division by $h$

$$u_x\left(1 - \frac{u^2}{hg}\right) + v_y\left(1 - \frac{v^2}{hg}\right) - \frac{uv}{hg}(u_y + v_x) = 0. \tag{2.10.48}$$

   Using (2.7.5) one obtains the potential equation

$$\varphi_{xx}\left(1 - \frac{\varphi_x^2}{gh}\right) + \varphi_{yy}\left(1 - \frac{\varphi_y^2}{gh}\right) - 2\varphi_{xy}\frac{\varphi_x\varphi_y}{gh} = 0, \tag{2.10.49}$$

   compare with (2.8.9)!

10. Derive the characteristics of (2.10.49):
    Hint: Using (1.2.1) one may read off $a, b, c$ from (1.2.7) and (2.10.49).

# 3. Water waves

## *3.1 The variety of water waves*

There exist many causes exciting waves in water: gravity, tide, wind, capillarity, earthquakes, underwater explosions, launching of rockets from submarines etc. Since wave propagation depends on the depth of the lake or ocean and on shore formation, there exist many words to describe the various wave types. We offer an alphabetic listing of wave names and give a short description of the phenomenon. (Please have in mind that terminology changes locally!) For the convenience of German speaking readers we include the German word [3.1]

1. *breaker* (Brecher, Brandungswelle), also waves of translation, form when waves enter shallow water. The rate of forward movement decreases and the wave height increases. Later on the wave falls back and becomes a breaker. Then a part of the water is thrown forward and does not return. Breaking especially occurs if the waves run against a submerged ridge. Breaking occurs when the wave amplitude is increased beyond a certain limit depending on the wave length and on the depth of water. The ratio wave height divided by water depth is the more important parameter in determining the non-breaking limit than the ratio height : length of the wave. See also short crested waves below.

2. *capillary waves, surface tension waves, ripples* (Oberflächenwelle, Riffel) are waves at the interface between two fluids or between air and fluid. Such ripples are generated by wind. They are of importance to the friction of air blowing over water. The wave height is a function of wind velocities. Ripples can be produced in a tank of water by a vibrating rod dipped just below the surface. If the water depth is more than half a ripple wave length the surface waves present a perfect analogy with a three-dimensional sound wave. Reducing the depth $h$ of the tank reduces the velocity of the ripples, giving the effect of a denser medium. This is an observational proof of the equivalence theorem, see also section 3.3.

3. *gravity waves* (Schwerewelle) are wind excited or due to tides. They are strongly dependent on the water depth, see section 3.2. They show dispersion: the wave propagation speed depends on the wave length $\lambda$. Waves with greater $\lambda$ propagate faster. If the water depth is very small

then there is no dispersion. If gravity waves and capillarity waves are considered together, the dispersion is more complicated, see section 3.3.

4. *internal waves, interfacial waves* (Raumwelle) are waves at the interface of two layers of water of different temperature and therefore density. The restoring force is mainly due to gravity. When light fluid from upper layers is depressed into the heavy lower layer, buoyance forces tend to return the layers to their equilibrium position. Observed internal waves can be analyzed into a spectrum with the shortest wave periods of a few minutes and longer periods up to many days. Internal waves have their maximum amplitude at the interface and diminish with distances above and below. Due to the small density difference of the two layers quite large amplitudes and instabilities (KELVIN-HELMHOLTZ *instability*) [2.8] may appear. The wave propagation speed is determined by gravity $g$, the density difference of the two layers and the thickness of the layers.
5. *jump, hydraulic jump, standing wave* (Wassersprung, Wasserstoß, stehende Welle). A wave that occurs where a stream of water changes from a *supercritical* condition to a *subcritical* condition is termed a hydraulic jump. It comprises an abrupt rise in water level through a region of intense turbulence. Standing jumps appear behind a localized irregularity (ridges etc) in an otherwise uniform flow channel.
6. *long water waves* (lange Welle, seichte Welle). These waves especially occur on the water surface with wave periods longer than 30 sec. There have been observed wave periods extending to about 120 minutes. These long waves are caused by many different effects. Those in the range 2–10 minutes are associated with the onset of heavy surf (see below) and are called surf beats. When the wave reaches the shore it is reflected back as a wave whose amplitude varies with the water depth. At any point some distance from the shore, the reflected wave which decreases in height only slowly with increasing water depth, is greater than the incident forced wave which increases more rapidly with decreasing depth. Long waves of periods 2–10 minutes can also be caused by wind and atmospheric disturbances but the period of waves caused by these effects are mainly greater and can be as 120 minutes.
7. *plumes, water columns* (Wasserfahne, Wassersäule) are not waves in the proper sense but effluents from a source. They are dispersed in

the atmosphere by the wind and stirring motions produced in clouds by the latent heat of vapor condensation. The energy of motion of a plume may be due to gravity or the CORIOLIS *force*.

8. *roll waves* (Roller, Rollwelle) occur in streams mainly very fast down a narrow channel like the spillway (Überlauf) of a dam, see also section 2.10. In the relevant literature [3.6] roll waves are defined as a type of water flow, which occurs in turbulent flows down inclined open channels, like spillways of dams, run-off channels etc. They have to be described by a nonlinear shallow water wave theory. The phenomenon is a two-dimensional irrotational flow of water over a rigid variable bottom of depth $-h(x)$, where the $x$-axis is taken horizontally and is identical with the direction of propagation. Gravity acts in the $z$-direction. If one designates by $\eta$ the vertical distance from the $x$-axis to the water surface, then $z = \eta + h > 0$ will measure the vertical height of the water surface above the inclined channel bottom. The equation describing these waves over a channel bed which is linear and inclined at an angle $\vartheta > 0$ below the horizontal then includes the given $h(x)$ and
$$\frac{\partial h(x)}{\partial x} = \tan \vartheta. \tag{3.1.1}$$
Assuming ("shallow water approximation") that the $z$-component of the water particle acceleration may be neglected, then the hydrostatic pressure is assumed to be given by
$$p = g\rho_0(\eta - z), \tag{3.1.2}$$
where $\rho_0$ is the constant water density. Then the equation of motion reads
$$u_t + uu_x = -g\eta_x \tag{3.1.3}$$
and the continuity equation becomes
$$\eta_t + \frac{\partial}{\partial x}(u[\eta + h]) = 0$$
or
$$u_x\eta + \eta_x u + u_x h + h_x u = -\eta_t. \tag{3.1.4}$$
Resisting forces may be added. *Progressing waves* like
$$u(x,t) = U(x - ct) = U(\xi) \tag{3.1.5}$$
offer solutions for waves and shocks [3.6].

9. *seiches* (Seiche, Binnenwasserwelle) are periodic fluctuations of the water level caused by disturbances similar to the sloshing back and forth of water in a large diskpan. In the ocean seiches occur in bays and lagoons, but also along the open coast. Their wave period may vary from minutes to hours and depends on the dimensions of the moving water masses. They may also be explained as lower frequency surface gravity waves. Some authors call long water waves of periods 15 - 35 minutes seiches. Seiches are commonly generated by wind, tides or oscillations of adjacent water bodies. Seiches may form standing waves in basins.

10. *shallow-water waves*, LAGRANGIAN *waves* (Seichtwasserwelle) are long water waves. They are both dispersive and refractive and are pure gravity waves in water with small depth $h$. The wave propagation velocity depends on both on the wave length $\lambda$ and the position in space. If the wave length is much less than the water depth $h$ the wave velocity becomes independent of position and dependent only on the wave length. No refraction occurs. In the extreme, if $\lambda \gg h$, the velocity becomes independent of $\lambda$ (no dispersion). Then the wave speed is given by $\sqrt{gh}$. This formula defines shallow-water waves, see section 3.2. If the variation of water depth is dominant, if it is a function of the horizontal coordinates $x, y$, then the wave speed depends on $x, y$. Characteristics and other graphical methods can describe the propagation. As the wave-fronts advance they always swing toward shallower water and away from deeper.
In the literature [3.7] shallow water waves are defined by the following hyperbolic partial differential equation of second order

$$\frac{\partial}{\partial x}(gh\varphi_x) + \frac{\partial}{\partial y}(gh\varphi_y) = \varphi_{tt}, \tag{3.1.6}$$

where $\varphi(x, y, t)$ is the wave amplitude potential, $g$ is the gravity acceleration and $h(x, y)$ is the variable water depth measured from the free water surface at rest. The wave amplitudes are very small ($\varphi_x^2 \approx 0$, $\varphi_y^2 \approx 0$).

11. *short crested waves* (kurzkämmige Welle). As deep ocean long waves enter shallow water, the wave length decreases and the wave becomes peaked or crested. For a critical height to wave length ratio, the water crest at the top moves faster than the crest. Thus the crest curls over and water flows on to the lower wave surface. Thus a doubly *modulated wave* appears. It is called a *breaker*.

12. *snoidal* (*cnoidal*) *waves* (Flachwasserwelle) are nonlinear periodic waves in shallow water. If they are symmetric around $x = 0$, they are cnoidal waves described by a JACOBI *function*. Cnoidal waves are also solutions of the *nonlinear* SCHRÖDINGER *equation* (1.5.20) and are dscribed by (1.5.24) [1.1]. In shallow water the ratio wave height divided by fluid depth is the more important parameter in determining the non-breaking limit than the ratio length to wave length. If the depth is constant and uniform the only possible periodic wave is the cnoidal wave.
13. *solitary waves, simple waves* (Einzelwelle) are the limits of cnoidal waves for infinite wave length. They travel at a speed $\sqrt{gh}$. They propagate without change of form with the amplitude depending on the velocity. *Solitons* adhere to this type of waves, see section 3.4.
14. *surf* (Brandung) is an expression for the piling up of waves entering into shallow water. The distortion of the water movement is associated with a decease of wave velocity and of wave length. Finally breakers appear.
15. *surface waves, surface gravity waves* (Oberflächenwelle, Schwerewelle) are excited on the water surface by wind, gravity and surface tension (capillarity). More details will be given in sections 3.2 and 3.3.
16. *surge, storm surges, flood waves* (Flutwelle, Sturzsee, Woge) is a sudden increase to an excessive or abnormal value of pressure or level in a fluid.
17. *swell* (waves) (Schwall, Dünung) is an expression for waves leaving their storm. Their wave length, velocity and period ($\approx 10$ s) increase as the waves travel away from the stormarea while their height decreases because of the resistance offered by the air. It is thus possible to judge the distance of a storm from the approaching swell. A short and high swell comes from a nearby storm. Each ocean region has its own characteristic swell related to the local atmospheric circulation.
18. *swirl* (Strudel, Wirbel) is the slow rotation of the flow of a fluid around an axis, see Fig. 2.2.
19. *tides* (Gezeiten, Ebbe und Flut) are alternating relative motions of water due to the gravitational forces of external bodies like the Moon and the Sun. *Tidal waves* (Gezeitenwellen) are generated by these astronomical bodies. A *tidal bore* (Springflutwelle) occurs if Moon and Sun and Earth are situated along a straight line.

20. *translation waves*, waves of translation (Transportwelle). If a wave profile becomes progressively distorted, the height of the crests above mean level becoming progressively larger than the depth of the troughs below. In such a situation water is therefore carried forward and the wave is called a translation wave. Due to velocity $\sim \sqrt{gh}$, the crests travel faster than the troughs and so in several wave lengths gain on them and the profile of the wave has sheer vertical forward face and then starts to break.

21. *trochoidal waves* (Trochoidalwelle) are waves with a trochoidal profile. In a gravity wave of finite amplitude the fluid elements (water particles) undergo a circular motion, provided the fluid is sufficiently deep and no interference occurs from the bottom. The amplitude of such *rotational waves* is equal to the diameter of the particles elementary circular path, the shape of the wave is trochoidal having sharp crests and smooth troughs,

22. *tsunamis* are large amplitude long nonlinear surface gravity waves of shock character formed by sudden dislocations of the ocean floor. They are NOT tidal waves or swells. Earthquakes of an intensity greater than seven on the RICHTER earthquake scale or underwater explosions may produce such shock waves in water. Their speed of propagation varies locally depending on the local depth of the ocean. It may be up to 900 km $h^{-1}$. The Indonesian tsunami of December 26, 2004 attained 750 km $h^{-1}$.decem The tsunami height in the open ocean is not very large but when tsunamis approach the shore their height increases with decreasing depth of the ocean and may reach a crest of 30 m. The tsunami waves diverge from their source over a circular wave front and due to spherical Earth surface, when they have traversed more than a quarter of the Earth's circumference, they tend to converge again to the antipode point of their sources. Thus for a past Chilean earthquake the tsunami effect on Japan has been many cases worse than that at the nearer situated Hawaii. To accurately model tsunami propagation over large distances, thus the Earth's curvature should be taken into account, including the CORIOLIS *force* and wave dispersion. The wave period of tsunamis may be between some minutes and one hour, their wave length may be up to hundreds of km. The rise of a tsunami being a shock wave in water is followed by a trough, a rarefaction wave. The arrival of a tsunami is first noticed by a fall in the ocean water level for several minutes, as if there were an abnormally low tide, but then followed by a rapid rise of the sea level by far exceeding the

highest tide level. Whereas sea and swell normally break and produce breakers, tsunamis come over the reefs and give rise to very turbulent water conditions with a high destructive power.
Up to now attempts at an exact theoretical analysis of tsunamis have met with only qualitatively success. The knowledge is more empirical. The reason for this situation is that shock waves like tsunamis may only be treated analytically, if two conditions are satisfied:

(a) inclusion of dissipative effects like viscosity which smooths the steepening and avoids infinite or double values,
(b) derivation of a (nonlinear) tsunami wave equation based on an equivalence principle as described in section 2.10.

The comparison of observed arrival time with model arrival time was however quite sufficient in the case of the Indonesian 2004 tsunami [3.3].
The life of a tsunami has been described in details by the Western Coastal and Marine Geology Institute [3.4]:

(a) Groundshaking due to earthquakes pushes the entire water column up. The potential energy resulting from pushing water above normal mean sea level is transferred to the kinetic energy of the horizontal propagation of the tsunami.
(b) Within several minutes of the ground shaking the initial tsunami is split into a tsunami that travels out to the deep ocean ("*distant tsunami*") and another tsunami that travels towards the nearby coast ("*local tsunami*"). The height of the original first tsunami is thus split into two halves. Since the tsunami propagation speed depends on the local water depth, the distant tsunami (deep ocean tsunami) travels faster than the local tsunami near shore.
(c) During propagation of the two tsunamis their amplitude increases and the wave length decreases when approaching a shore. In both cases a steepening of the wave occurs.
(d) As a tsunami wave travels from deep-water to a shore, runup occurs. *Runup* is a measurement of the height of the water ashore observed above the reference sea level. Tsunamis will often travel much farther inland than normal waves.
(e) After runup, part of the tsunami energy is reflected back to the open ocean. Additionally a tsunami can generate an *edge wave* (Randwelle, Kantenwelle) that travels back and forth parallel to

the shore. Many arrivals of a tsunami at a particular coast may occur. Even the first runup of a tsunami is often not the largest – several hours after the first arrival another tsunami may arrive which is larger than the first one. Seen the influences of local pecularities ("boundary conditions") it is very doubtful if an exact mathematical description of tsunamis is possible.

## Problems

1. A trochoid is a curve defined by the parametric representation

$$x = a(t - \lambda \sin t), \quad y = a(1 - \lambda \cos t). \tag{3.1.7}$$

Here $a$ is the radius of a rolling circle and $\lambda$ a parameter. Hint: use $a = 0.8$, $\lambda = 1.$, and the commands

```
x[t_]=a*(t-λ*Sin[t];y[t_]=a*(1-λ*Cos[t]); a=0.8;λ=1.;
ParametricPlot[{x[t],y[t]},{t,0,4*Pi}]
```

to produce Fig. 3.1.

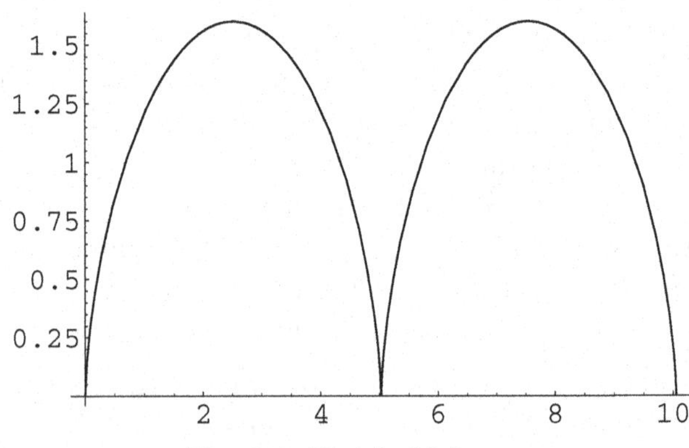


Fig. 3.1. Trochoidal wave

2. Derive the characteristics for the system (3.1.3), (3.1.4). Hint: Insert $\eta(x, t) = z(x, t) - h(x)$ into (3.1.3) and (3.1.4), use (1.4.18) and (1.4.23). Remember: $h(x)$ is a given function.

Solution: $R = 0$ gives

$$\frac{\mathrm{d}x}{\mathrm{d}t} = u \pm \sqrt{gh}. \tag{3.1.8}$$

3. Use progressing waves

$$u(x,t) = U(x-ct) = U(\zeta), \quad z(x,t) = Z(x-ct) = Z(\zeta) \tag{3.1.9}$$

to solve

$$u_t + uu_x + gz_x = gh_x, \tag{3.1.10}$$

$$z_t + zu_x + uz_x = 0, \tag{3.1.11}$$

where $h(x)$ is a given function and $c$ is assumed to be const. Calculate $U'$ and $Z'$. Explain the (fundamental!) difference between (3.1.10), (3.1.11) and (2.10.39), (2.10.40). Is it possible to derive an ordinary differential equation for $U(\zeta)$ or $Z(\zeta)$?

4. *Viscous waves* in viscous gases or liquids are propagated by virtue of the transverse shear reactions due to viscosity. They are thus transverse waves. Study the relevant literature, [2.2, pp. 3–47].

## *3.2 Gravity water waves*

Whereas vortices transport fluid elements, waves transport energy (and phase). It is important to understand the influence of the actual ocean depth $h$ on the propagation of water waves. This point is fundamental for a clear understanding of tsunamis. Gravity waves on the water surface are excited by wind or other actions disturbing the water surface at rest. In this section we assume that viscosity, heat conduction and capillarity effects may be neglected. If the CORIOLIS force is not taken into account, then gravity alone can propagate disturbances. Assuming that we consider a plane lake and that gravity acts in the $z$-direction, then we may write

$$V = \rho_0 g z \tag{3.2.1}$$

for the gravity potential energy $V$. $\rho_0$ is the constant water density. Assuming a three-dimensional potential flow (2.7.16), the equation of motion (2.3.20) may be written

$$\nabla\left(\frac{\partial\varphi}{\partial t}\rho_0 + \rho_0\frac{\vec{v}^2}{2} + p + \rho_0 g z\right) = 0. \tag{3.2.2}$$

Integration yields the *time dependent* BERNOULLI *equation*

$$\frac{\partial\varphi}{\partial t} + \frac{\vec{v}^2}{2} + \frac{p}{\rho_0} + gz = F(t). \tag{3.2.3}$$

This is now the boundary condition for the potential equation (2.7.6). Neglecting nonlinear terms and the integration constant $F(t)$, one obtains

$$\frac{\partial \varphi}{\partial t} + \frac{p}{\rho_0} + gz = 0. \tag{3.2.4}$$

The potential $\varphi$ depends on $x, y, z, t$. On the water surface defined by $z = 0$, the pressure $p_0$ is constant (1 at). Thus (3.2.4) may be written

$$\frac{\partial \varphi}{\partial t} = -gz + \text{const}, \quad \text{for} \quad z = 0, \tag{3.2.5}$$

where $p_0/\rho_0$ has been incorporated into const. The normal component $v_n$ of the water particles must vanish on the bottom of the lake or ocean, defined by $-z = h$

$$\frac{\partial \varphi}{\partial z} = v_n = 0 \quad for \quad z = -h, \tag{3.2.6}$$

see (2.7.18). This equation is the *bottom boundary condition.*

Assuming again $\mathrm{d}/\mathrm{d}t \approx \partial/\partial t$, derivation of (3.2.5) with respect to time yields

$$\frac{\partial^2 \varphi}{\partial t^2} = -g\frac{\partial z}{\partial t} = -gv_z = -g\frac{\partial \varphi}{\partial z}. \tag{3.2.7}$$

Additionally (2.7.6) holds. In order to solve these two equations we setup

$$\varphi(x, y, z, t) = \psi(x, y)U(z)\exp(i\omega t). \tag{3.2.8}$$

Insertion into (3.2.7) yields

$$\omega^2 U = gU_z, \tag{3.2.9}$$

which plays the role of a boundary condition for the second equation (2.7.6) $\Delta\varphi = 0$ which now reads

$$\frac{1}{\psi}(\psi_{xx} + \psi_{yy}) + \frac{U_{zz}}{U} = 0. \tag{3.2.10}$$

Designating by $k$ the separation constant, we obtain from (3.2.10) these two equations

$$\psi_{xx} + \psi_{yy} + k^2\psi = 0, \tag{3.2.11}$$

$$U_{zz} - k^2 U = 0. \tag{3.2.12}$$

The solution of (3.2.11) is

$$\psi(x, y) = A\exp i(k_x x + k_y y), \tag{3.2.13}$$

where $k_x^2 + k_y^2 = k^2$ designates the *wave vector* $\vec{k}$. The solution of (3.2.12) is given by

$$U(z) = \cosh k(z+h), \tag{3.2.14}$$

where $h$ is an integration constant. Now (3.2.6) is satisfied for $z = -h$

$$U_z(-h) = k \sinh k(z+h) = 0. \tag{3.2.15}$$

The variable $h$ designates now the variable lake or ocean depth. The solution (3.2.14) has also to satisfy (3.2.9). Insertion of (3.2.14) and of $U_z$ into (3.2.9) gives

$$\omega^2 = gk \tanh k(z+h), \tag{3.2.16}$$

so that for the waves on the surface ($z = 0$) the *dispersion relation*

$$\omega(k) = \sqrt{gk \tanh(kh)} = \sqrt{g\frac{2\pi}{\lambda} \tanh\left(2\pi\frac{h}{\lambda}\right)} \tag{3.2.17}$$

appears. Here $k = 2\pi/\lambda$ has been used. One now can find the following statements for these gravitational waves:

1. The surfaces of constant phase are given by the planes

$$k_x x + k_y y = \text{const.} \tag{3.2.18}$$

2. The surfaces of constant amplitude are defined by the planes

$$z = \text{const.} \tag{3.2.19}$$

3. If the surfaces of constant phase and of constant amplitude do not coincide, waves are called inhomogeneous. Thus the gravitational water waves are *inhomogeneous waves.*

4. The wave propagation speed $c$ depends on the wave length $\lambda$: waves with larger $\lambda$ propagate faster than waves with smaller $\lambda$ (*normal dispersion*, $\mathrm{d}c/\mathrm{d}\lambda > 0$). Thus a steepening up may occur, if various $\lambda$ are involved.

In order to prove this statement, we use (1.1.4) for the phase propagation speed $c$ giving now

$$c^2 = \frac{\omega^2(k)}{k^2} = \frac{\lambda^2\omega^2(k)}{(2\pi)^2} = \frac{\lambda g}{2\pi}\tanh\left(2\pi\frac{h}{\lambda}\right). \tag{3.2.20}$$

Three cases are of interest:

1. very deep water: $h \gg \lambda$. Then one has

$$c = \sqrt{(\lambda g)/(2\pi)}, \tag{3.2.21}$$

since $\tanh(\infty) = 1$. [1.4], [1.5]. This is a proof for statement 4. We have normal dispersion according to (3.2.21).

2. shallow water: $h \ll \lambda$. Then one may approximate tanh by its argument, $\tanh(2\pi(h/\lambda)) \approx 2\pi(h/\lambda)$. [1.4], [1.5]. Then (3.2.20) becomes

$$c = \sqrt{gh}, \tag{3.2.22}$$

compare with (2.3.22), which however describes quite another situation. (3.2.22) states that in shallow water no dispersion occurs. Thus the effect of steepening and turning over of the wave should take place in a nonlinear wave.

3. deep water: $h \approx \lambda$. Then we have dispersion according to (3.2.17) and (3.2.20).

A plot of (3.2.20) generated by the Mathematica command

```
Clear[f,c];h=50000.0,c0=Sqrt[h]
c[λ_]=Sqrt[λ*Tanh[h/λ]];
Plot[c[λ],{λ,0,5.*h}]
```

may give a clear picture, see Fig. 3.2.

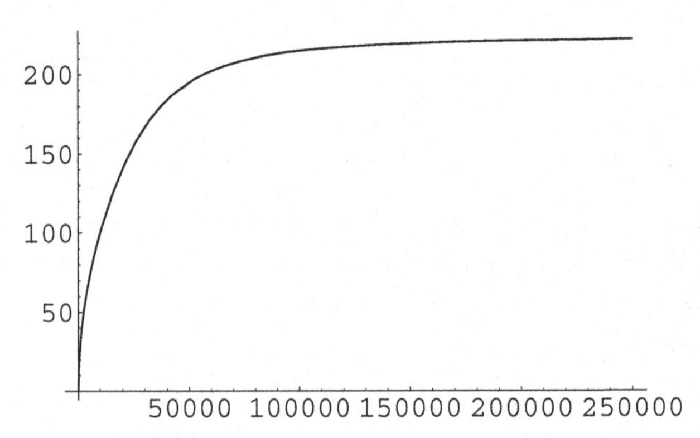


Fig. 3.2. Dispersion relation of gravitational waves

If there is no dispersion then the propagation speed is equal to the phase speed $c$. In the case of gravity water wave, the propagation speed of the wave power is equal to the group velocity $c_g$, since there is dispersion. The group velocity is given by (1.1.9). If we insert (3.2.17) into (1.1.9) one obtains

$$c_g = \frac{\mathrm{d}\omega(k)}{\mathrm{d}k} = \frac{ghk\,\mathrm{sech}^2(hk) + g\tanh(hk)}{2\sqrt{gk\tanh(hk)}}. \tag{3.2.23}$$

This result has been obtained by the Mathematica command

```
D[Sqrt[g*k*Tanh[k*h]],k]
```

Up to now we had assumed that the bottom of the ocean is plane, so that the distance water surface to bottom is constant.

Now we consider the case of an ascending bottom as it is the case near the shore. Then $h$ is small and for a decrease of the water depth from $h_a$ down to $h_b$ (3.2.22) yields

$$\frac{c_b}{c_a} = \sqrt{\frac{h_b}{h_a}} > 1. \tag{3.2.24}$$

This indicates that the wave crest characterized by $h_a$ is propagating faster ($c_b > c_a$) than the wave trough characterized by $h_a$ and the speed $c_a$, see also problem 2 in section 3.1.

The bottom boundary condition (3.2.6) forces fluid elements to oscillate on the bottom within the $x, y$ plane at $z = -h$. If the water is at rest, the coordinates of such a fluid element are given by $x, y, z = -h$. If a perturbation occurs due to a wave, the coordinates may be $x', y', z = -h$. Thus then a displacement $\xi, \eta$ occurs.

$$\xi = x' - x, \quad \eta = y' - y. \tag{3.2.25}$$

Since the particle speed is given by $\nabla\varphi$, we may write

$$v_x = \frac{\mathrm{d}\xi}{\mathrm{d}t} = \frac{\partial\varphi}{\partial x}, \quad v_y = \frac{\mathrm{d}\eta}{\mathrm{d}t} = \frac{\partial\varphi}{\partial y}. \tag{3.2.26}$$

The difference between $\xi, \eta$ and $x, y$ is of higher order and has been neglected. Then we have after integration with respect to $t$

$$\xi = C\frac{k}{\omega}\exp(ikx - i\omega t)\cosh k(z+h),$$

$$\eta = iC\frac{k}{\omega}\exp(ikx - i\omega t)\sinh k(z+h). \tag{3.2.27}$$

Now we use the abbreviations

$$a = \frac{Ck}{\omega}\cosh k(z+h), \quad b = \frac{Ck}{\omega}\sinh k(z+h) \tag{3.2.28}$$

and rewrite (3.2.27)

$$\xi = a\cos(kx - \omega t), \quad \eta = -b\sin(kx - \omega t). \tag{3.2.29}$$

Elimination of $t$ yields the equation of an ellipse

$$\frac{\xi^2}{a^2} + \frac{\eta^2}{b^2} = 1. \tag{3.2.30}$$

The ratio of the semiminor to the semimajor axes is given by

$$\frac{b}{a} = \tanh k(z+h) = \begin{matrix} 0 & \text{bottom} & z = -h \\ \tanh kh & \text{surface} & z = 0. \end{matrix} \tag{3.2.31}$$

For very deep water ($h \gg \lambda$) one has $b/a = \tanh(\infty) = 1$ and the ellipse becomes a circle which shrinks to a point at $z = -h$ due to (3.2.15). For shallow water ($h \ll \lambda$), the semiminor axis $b$ shrinks to zero due to $kh \to 0$, $k(z+h) \to 0$ and $b/a = \tanh(0) = 0$. Then the particles oscillate along a line of the length $2a$.

If the bottom is not a horizontal plane, then the *bottom boundary condition* depends on the solution $\varphi(x, y, z, t)$. The actual location $x, y$ of a wave on the water surface $z = 0$ is described by this solution, but the solution itself depends on the boundary condition: we have a *nonlinear boundary problem* [1.1], because during the propagation of the wave the maximum value of $h$, namely $h_0$ is no longer constant but varies. It depends on $x, y, t$. One has a *moving boundary problem* which may be solved using similarity transformations.

A boundary problem of a linear differential equation may become nonlinear, if the boundary condition itself is nonlinear. If one does not neglect the nonlinear terms in (3.2.3), then the BERNOULLI equation reads

$$\frac{\partial\varphi}{\partial t} + \frac{(\nabla\varphi)^2}{2} + \frac{p}{\rho_0} + gz = 0. \tag{3.2.32}$$

On the water surface $z \equiv 0$ one has

$$\frac{\partial\varphi}{\partial t} + \frac{(\nabla\varphi)^2}{2} + gz = \text{const}, \tag{3.2.33}$$

see (3.2.5), whereas on the bottom $z = -h_0$ one has again

$$\left(\frac{\partial\varphi}{\partial z}\right)_{z=-h} = v_n = 0, \tag{3.2.34}$$

see (3.2.6). Integration of $\Delta\varphi = 0$ together with the two boundary conditions (3.2.33) and (3.2.34) results in *cnoidal waves*, see (1.5.24).

Moving boundaries are mainly nonlinear, although very often it is difficult to recognize the nonlinearity. Consider an infinite half-space orthogonal to the $x$ axis, which is filled with water. On the boundary plane $x = 0$ the temperature is cooled down to $T_0 < 0°$ C. Thus the water will freeze there. As time passes on, a *freezing front* will penetrate into the water and freeze it. If we designate by $T_1(x,t)$ the temperature within the ice and by $T_2(x,t)$ within the water and let $T_s$ be the temperature within the front, then continuity demands

$$T_s = T_1(x = X(t), t) = T_2(x = X(t), t) \quad \text{or} \quad \mathrm{d}T_1 = \mathrm{d}T_2. \tag{3.2.35}$$

Here $X(t)$ designates the actual location of the freezing front. Now we consider the two heat conduction equations. Subscript 1 refers to ice and subscript 2 to water, respectively:

$$\frac{\partial T_1(x,t)}{\partial t} = \frac{\lambda_1}{\rho_1 c_1}\frac{\partial^2 T_1(x,t)}{\partial x^2}, \qquad \frac{\partial T_2(x,t)}{\partial t} = \frac{\lambda_2}{\rho_2 c_2}\frac{\partial^2 T_2(x,t)}{\partial x^2}, \tag{3.2.36}$$

where $\rho$ [kg/dm$^3$] and $c$ [kJ/kg K] designate density and specific heat, K is the thermodynamic temperature, $\lambda$ [W/mK] is the *thermal conductivity* (heat conductivity). Boundary conditions exist mainly within the freezing front. Energy conservation demands that the relative rate of heat flow through the front is equal to the melting heat $L$ [kJ/kg] transported away [1.1]:

$$\lambda_1 \frac{\partial T_1(x = X(t), t)}{\partial t} - \lambda_2 \frac{\partial T_2(x = X(t), t)}{\partial t} = L\rho\frac{\mathrm{d}X}{\mathrm{d}t}, \tag{3.2.37}$$

where $\rho = (\rho_1 + \rho_2)/2$ (Stefan *boundary condition*). Furthermore, the conditions at $x = 0$ and $x = \infty$ have to be taken into account:

$$T_1(0,t) = T_0, \quad T_2(\infty, t) = T^*, \tag{3.2.38}$$

where $T^*$ is the initial water temperature:

$$T_2(x,0) = T^* \quad \text{for} \quad x \geq 0. \tag{3.2.39}$$

Having a short look at the foregoing equations, it is not immediately clear that they represent a nonlinear problem. But using

$$\mathrm{d}T_i = \frac{\partial T_i}{\partial t}\mathrm{d}t + \frac{\partial T_i}{\partial x}\mathrm{d}x \quad \text{for} \quad i = 1, 2 \tag{3.2.40}$$

and

$$\frac{\mathrm{d}x}{\mathrm{d}t} = \frac{\mathrm{d}X}{\mathrm{d}t} = \left(\frac{\partial T_1}{\partial t} - \frac{\partial T_2}{\partial t}\right) \Big/ \left(\frac{\partial T_2}{\partial X} - \frac{\partial T_1}{\partial X}\right) \tag{3.2.41}$$

and insertion into (3.2.37) yields the *nonlinear boundary condition*

$$\left(\frac{\partial T_2}{\partial X} - \frac{\partial T_1}{\partial X}\right)\left(\lambda_1 \frac{\partial T_1}{\partial X} - \lambda_2 \frac{\partial T_2}{\partial X}\right) = L\rho\left(\frac{\partial T_1}{\partial t} - \frac{\partial T_2}{\partial t}\right). \tag{3.2.42}$$

The nonlinearity is hidden in the fact that the moving front boundary $X(t)$ depends on the temperature $T(x,t)$.

## Problems

1. Solve the STEFAN problem (3.2.36). Use the similarity transformation

$$\eta = x/2\sqrt{\tau}, \quad \tau = \lambda t/\rho c, \quad a = \lambda/\rho c \tag{3.2.43}$$

giving

$$\frac{\mathrm{d}^2 T(\eta)}{\mathrm{d}\eta^2} \equiv -2\eta \frac{\mathrm{d}T(\eta)}{\mathrm{d}\eta}. \tag{3.2.44}$$

Hints: $T'(\eta) = u(\eta)$ gives

$$T(\eta) = \mathrm{const} \int_0^\eta \exp(-\eta^2)\mathrm{d}\eta = T_0 + \mathrm{erf}\left(\frac{x}{2\sqrt{at}}\right).$$

Define the path of the freezing front $X(t) = \alpha t^{1/2}$, calculate $\mathrm{d}X/\mathrm{d}t$ and satisfy the interface conditions at $X(t)$.

$$\left.\frac{\partial T_1(x,t)}{\partial x}\right|_{x=X(t)} \quad \text{and} \quad \left.\frac{\partial T_2(x,t)}{\partial x}\right|_{x=X(t)}.$$

Solution [1.1]: (for erf and erfc, see (3.2.64) below)

$$T_1(x,t) = T_0 + \frac{T_s - T_0}{\mathrm{erf}\,(\alpha/2\sqrt{a_1})}\mathrm{erf}\left(\frac{x}{2\sqrt{a_1 t}}\right), \quad 0 \le x \le X(t),$$

$$T_2(x,t) = T^* + \frac{T_s - T^*}{\mathrm{erfc}\,(\alpha/2\sqrt{a_2})}\mathrm{erfc}\left(\frac{x}{2\sqrt{a_2 t}}\right), \quad x \ge X(t).$$

2. Prove that the *shallow water wave equation* (3.1.5) is a linear approximation $(\nabla\varphi)^2 \approx 0$. Hints: Write (2.10.39), (2.10.40) in three-dimensional form, use $\vec{v} = -\nabla\varphi$ and (2.3.18) as well as $\mathrm{curl}\,\vec{v} = 0$.

Derive the potential equation for $\varphi(x, y, z, t)$. The time dependent BERNOULLI equation

$$-\nabla\frac{\partial\varphi}{\partial t} + \frac{1}{2}\nabla(\nabla\varphi)^2 + \nabla hg = 0 \tag{3.2.45}$$

delivers $h$ after integration over space

$$-\frac{\partial\varphi}{\partial t} + \frac{1}{2}(\nabla\varphi)^2 + hg = 0, \tag{3.2.46}$$

and after derivation with respect to time one obtains $(\partial h)/(\partial t)$ from

$$-\frac{\partial^2\varphi}{\partial t^2} + \frac{1}{2}\frac{\partial}{\partial t}(\nabla\varphi)^2 + \frac{\partial hg}{\partial t} = 0. \tag{3.2.47}$$

Inserting $gh_t$ from (3.2.47) into the continuity equation

$$\frac{\partial h}{\partial t} - \operatorname{div}(h\nabla\varphi) = 0 \tag{3.2.48}$$

yields

$$\varphi_{tt} = \operatorname{div}(gh\nabla\varphi) + \frac{1}{2}\frac{\partial}{\partial t}(\nabla\varphi)^2. \tag{3.2.49}$$

3. Investigate the effect of viscosity on the equation of motion. Use

$$\frac{\partial u}{\partial t} + u\frac{\partial u}{\partial x} = \nu\frac{\partial^2 u}{\partial x^2}. \tag{3.2.50}$$

This equation (and also (3.2.52) [2.13]) is called BURGERS *equation.* This nonlinear equation may be linearized. For this purpose use a HOPF *transformation*

$$u = -2\nu\frac{\partial}{\partial x}\ln\psi. \tag{3.2.51}$$

Hint: transform $u = v_x$ to obtain

$$v_{xt} + v_x v_{xx} = \nu v_{xxx} \tag{3.2.52}$$

and integrate with respect to $x$ giving

$$v_t + \frac{1}{2}v_x^2 = \nu v_{xx}. \tag{3.2.53}$$

Then apply the HOPF transformation in the form

$$v = -2\nu\ln\psi. \tag{3.2.54}$$

This gives the solution $\nu\psi_{xx} = \psi_t$.

4. Investigate a two-dimensional steady water flow over a plane bottom [2.21]. Let $h(x, y)$ be the distance between the (variable) water surface and the bottom $z = 0$. The local hydrostatic pressure $z$ is determined by the vertical distance $h - z$, compare (2.10.9)

$$p = \rho_0 g(h - z). \tag{3.2.55}$$

Gravity acts in the vertical direction ($z$-direction). Neglecting vertical acceleration of fluid elements in relation to gravity, one obtains

$$\frac{\partial p}{\partial x} = \rho_0 g \cdot \frac{\partial h}{\partial x}, \quad \frac{\partial p}{\partial y} = \rho_0 g \cdot \frac{\partial h}{\partial y}. \tag{3.2.56}$$

Using the equivalence principle of section 2.10, the continuity equation can be written

$$\frac{\partial(hu)}{\partial x} + \frac{\partial(hv)}{\partial y} = 0. \tag{3.2.57}$$

Using $u = \varphi_x, v = \varphi_y$ derive the potential equation for this flow.

Solution:

$$\varphi_{xx}\left(1 - \varphi_x^2/gh\right) + \varphi_{yy}\left(1 - \varphi_y^2/gh\right) - 2\varphi_x\varphi_y\varphi_{xy}/gh = 0. \tag{3.2.58}$$

5. To investigate the BURGERS equation, solve

$$\nu\psi_{xx} = \psi_t. \tag{3.2.59}$$

This may be done by a LAPLACE *transformation* according to p. 104 in [1.1] or by a *separation setup*

$$\psi(x, t) = X(x) \cdot T(t) \tag{3.2.60}$$

yielding two ordinary differential equations. This could be done by the Mathematica command ([1.1, p. 99]):

```
ψ[x_,t_]=X[x]*T[t]
Expand[(ν*D[ψ[x,t],{x,2}]-D[ψ[x,t],t])/(X[x]*T[t])]
```
(3.2.61)

Another way to solve (3.2.59) would be by a *similarity transformation* [1.1]

$$\psi(x, t) = F(\eta), \quad \eta = x^\alpha t^\beta. \tag{3.2.62}$$

Inserting into (3.2.59) yields an ordinary differential equation for $F(\eta)$ which reads

$$\nu F''\alpha^2\eta^2 + F'\alpha(\alpha-1)\eta + F'\beta\eta x^2/t = 0. \qquad (3.2.63)$$

Hint: assume $x/\sqrt{t} = \eta$ which gives $\alpha = 1$, $\beta = -1/2$. Remark: the following designation is usual

$$\frac{2}{\sqrt{\pi}}\int_0^x \exp(-\xi^2)\mathrm{d}\xi = \operatorname{erf}(x), \quad \operatorname{erfc}(x) = 1 - \operatorname{erf}(x). \qquad (3.2.64)$$

The function erf is called *error function* [1.4].

## *3.3 Capillarity waves*

Surface tension or capillarity is the force acting in the surface of a liquid, tending to minimize the surface area. Whereas within the body of the liquid the time-averaged force exerted on any given molecule by its neighbors is zero, molecules on the surface have no neighbors. Beyond the free surface, there exist no molecules to counteract the forces of attraction exerted by molecules in the interior. These effects generate a *surface energy* $U$ defined by

$$U = \sigma F, \qquad (3.3.1)$$

where $F$ is the area in $\mathrm{cm}^2$ and $\sigma$ is the capillarity constant in $\mathrm{erg\,cm}^{-2}$. The surface tension is not constant, it depends on temperature. For water it varies between 75.6 at 0° C and 71.18 $\mathrm{erg\,cm}^{-2}$ at 30° C. For a local displacement $u(x, y, t)$ of the water surface due to perturbation by wind, one finds the pressure [2.2], [2.10]

$$p = -\sigma\left(\frac{\partial^2 u}{\partial x^2} + \frac{\partial^2 u}{\partial y^2}\right). \qquad (3.3.2)$$

Since equation (2.7.6) $\Delta\varphi = 0$ is still valid for the potential $\varphi(x, y, z, t)$ of the water surface displacement, the solution given in section 3.2 is valid, too:

$$\varphi(x, y, z, t) = A\exp\, i(k_x x + k_y y)\cosh\, k(z+h)\exp(i\omega t). \qquad (3.3.3)$$

The boundary condition is again given by the time-dependent BERNOULLI equation (3.2.4). For $g = 0$, it reads

$$\frac{\partial\varphi}{\partial t} - \frac{1}{\rho_0}\sigma\left(\frac{\partial^2 u}{\partial x^2} + \frac{\partial^2 u}{\partial y^2}\right) = 0. \qquad (3.3.4)$$

Derivation with respect to time and the replacement $u_t \to \varphi_z$ yields [2.10]

$$\frac{\partial^2 \varphi}{\partial t^2} - \frac{\sigma}{\rho_0} \frac{\partial}{\partial z} \left( \frac{\partial^2 \varphi}{\partial x^2} + \frac{\partial^2 \varphi}{\partial y^2} \right) = 0 \quad \text{for} \quad z = 0. \tag{3.3.5}$$

This represents the boundary condition on the water surface for $\varphi$ defined by (2.7.6). A solution of (2.7.6) is given by

$$\varphi = A \exp(kz) \cos(kx - \omega t). \tag{3.3.6}$$

Insertion into the boundary condition delivers the dispersion relation

$$\omega^2 = \frac{\sigma}{\rho_0} k^3 \tag{3.3.7}$$

or

$$c = \sqrt{\frac{\sigma k}{\rho_0}} = \sqrt{\frac{\sigma}{\rho_0} \frac{2\pi}{\lambda}}. \tag{3.3.8}$$

Waves defined by (3.3.8) are called capillary waves or *ripples* (Riffeln, Kräuselwellen). Equation (3.3.8) indicates *anomalous dispersion*: waves with small wave-lengths are faster ($\mathrm{d}c/\mathrm{d}\lambda < 0$).

In the foregoing calculations gravity has been neglected, but gravity is always present. Thus we add the term $g(\partial\varphi/\partial z)$ on the lhs of equation (3.3.5), so that the boundary condition now reads

$$g\frac{\partial \varphi}{\partial z} + \frac{\partial^2 \varphi}{\partial t^2} - \frac{\sigma}{\rho_0} \left( \frac{\partial^2 \varphi}{\partial x^2} + \frac{\partial^2 \varphi}{\partial y^2} \right) = 0 \quad \text{for} \quad z = 0. \tag{3.3.9}$$

Repeating the calculation, one now gets the more complicated dispersion relation

$$\omega^2 = gk + \frac{\sigma}{\rho_o} k^3 = g\frac{2\pi}{\lambda} + \frac{\sigma}{\rho_0} \left( \frac{2\pi}{\lambda} \right)^3. \tag{3.3.10}$$

Thus the phase speed is given by

$$c = \frac{\omega}{k} = \sqrt{\frac{g}{k} + \frac{\sigma k}{\rho_0}} = \sqrt{\frac{g\lambda}{2\pi} + \frac{\sigma}{\rho_0} \frac{2\pi}{\lambda}}. \tag{3.3.11}$$

To investigate the combined effect of gravity and capillarity and the new dispersion relation we calculate the minimum of $c(\lambda)$. The derivation $\mathrm{d}c/\mathrm{d}\lambda = 0$ yields the minimum of $c$ at

$$\lambda_{\text{min}} = 2\pi \sqrt{\frac{\sigma}{\rho_0 g}} \tag{3.3.12}$$

and

$$c_{\text{min}}^2 = 2\sqrt{\frac{\sigma}{\rho_0 g}}. \tag{3.3.13}$$

These results have the following meaning:

1. No waves exist which propagate slower than with $c_{\text{min}}$.
2. All waves, $\lambda > \lambda_{\text{min}}, \lambda = \lambda_{\text{min}}$ and $\lambda < \lambda_{\text{min}}$ propagate with a speed larger than $c_{\text{min}}$.

Waves $\lambda < \lambda_{\text{min}}$ are usually called ripples and waves $\lambda > \lambda_{\text{min}}$ are gravitational waves, sometimes accompanied by ripples on their flanks.

**Problems**

1. Calculate $\lambda_{\text{min}}$ and $c_{\text{min}}$.

   Solution: $\lambda_{\text{min}} \approx 0.017\,\text{m}$, $c_{\text{min}} \approx 23\,\text{cm}\,\text{s}^{-1}$.

2. Show that the phase speed (propagation speed) is equal to the group velocity if there is no dispersion. Show that the group velocity is smaller than the phase speed if there is normal dispersion. For $\lambda < \lambda_{\text{min}}$ one has anomalous dispersion. What happens for $\lambda = \lambda_{\text{min}}$?

3. Calculate the group velocity from (3.3.11).

4. *Anual waves* are generated on the surface of a lake if a stone is thrown into the water. Calculate the velocity potential $\varphi(r, \vartheta, z)$ in cylindrical coordinates. Hint: Solve

$$\varphi_{rr} + \frac{1}{r}\varphi_r + \frac{1}{r^2}\varphi_{\vartheta\vartheta} + \varphi_{zz} = 0. \tag{3.3.14}$$

   Assume $\partial/\partial\vartheta = 0$.

   Solution:

$$\varphi(r, z, t) = A\exp(-kz)\exp(-i\omega t)f(r).$$

   Show that $f(r)$ is given by BESSEL functions and $\omega = \sqrt{gh}$.

## *3.4 Solitons*

A soliton wave or *soliton* is a nonlinear water wave which does not change its shape during propagation. Such waves in channels are well known since long time [3.8] [3.9]. The two effects: nonlinearity and dispersion seem to compensate each other, so that no physical cause for steepening or modifying the shape of these nonlinear waves exists. Solitons are a typical nonlinear phenomenon – linearized hydrodynamic equations never have soliton solutions. Their stability can, to some extent, be understood as a consequence of dispersion and nonlinearity. As we have seen in chapter 3.2, long water waves travel faster than short waves. At the same time nonlinear effects lead to a concentration of the original pulse.

It is usual to describe solitons as solutions of the KORTEWEG-DE VRIES equation which we shall derive now [1.1]. The potential function $\varphi(x,z,t)$ satisfies

$$\Delta\varphi = 0 \tag{3.4.1}$$

even for a nonlinear but non dissipative wave like solitons. The nonlinearity comes in from the boundary conditions. We consider a wave propagating in the $x$-direction over a lake or channel surface of constant depth $h_0$. Let $h(x,z,t)$ describe the disturbed water surface. The BERNOULLI *equation* (3.2.3) will be used in the form

$$\varphi_t + \frac{1}{2}\left(\varphi_x^2 + \varphi_z^2\right) + g(h - h_0) = 0. \tag{3.4.2}$$

If $z = 0$ designates the bottom at the channel – in contradiction to previous designations – the normal velocity component $v_n$ vanishes at $z = 0$

$$v_n = -\frac{\partial\varphi}{\partial z} = 0, \quad \text{for} \quad z = 0. \tag{3.4.3}$$

This is now the bottom boundary condition. It is still linear. On the free water surface now defined by $z = h(x,z,t)$ the *free boundary condition* is however nonlinear. It reads

$$-v_n = \varphi_z = \frac{\mathrm{d}h}{\mathrm{d}t} = h_t + \varphi_x h_x, \quad \text{for} \quad z = h(x,z,t). \tag{3.4.4}$$

Dissipation has been neglected. Some authors add terms describing wind pressure or capillary tension.

We shall now solve the two differential equations (3.4.1) for $\varphi(x,z,t)$ and (3.4.2) for $h(x,z,t)$ taking into consideration the two boundary conditions (3.4.3) and (3.4.4). We introduce *stretched variables* $\xi, \tau, \psi$ by

$$\xi = \sqrt{\varepsilon}(x - c_0 t), \quad \tau = \varepsilon^{3/2} t, \quad \psi(\xi,z,\tau) = \sqrt{\varepsilon}\varphi(x,z,t). \tag{3.4.5}$$

Here $c_0$ is an abbreviation and $\varepsilon$ a small parameter.

$$c_0 = \sqrt{gh_0}, \tag{3.4.6}$$

compare (3.2.22)! Introduction of these variables into (3.4.1), (3.4.2), (3.4.3) and (3.4.4) delivers

$$\varepsilon\psi_{\xi\xi} + \psi_{zz} = 0, \quad \psi_z = 0, \quad \text{for} \quad z = 0, \tag{3.4.7}$$

$$\psi_z = \varepsilon^2 h_\tau + \varepsilon(\psi_\xi - c_0)h_\xi, \quad \text{for} \quad z = h, \tag{3.4.8}$$

$$\varepsilon^2\psi_\tau - \varepsilon c_0\psi_\xi + \frac{1}{2}\left(\varepsilon\psi_\xi^2 + \psi_z^2\right) + \varepsilon g(h - h_0) = 0. \tag{3.4.9}$$

Using $\varepsilon$ we now write down a perturbation setup for the two unknown functions

$$h = h_0 + \varepsilon h_1 + \varepsilon^2 h_2, \quad \psi = \varepsilon\psi_1 + \varepsilon^2\psi_2. \tag{3.4.10}$$

Then one gets

$$\begin{aligned}
&\psi_{1zz} = 0, \ \ \psi_{nzz} + \psi_{n-1\xi\xi} = 0,\\
&\psi_1 = \psi_1(\xi,\tau), \quad \psi_2 = -z^2\psi_{1\xi\xi}/2, \quad \psi_3 = z^4\psi_{1\xi\xi\xi\xi}/24,\\
&\psi_{2z} = -c_0 h_{1\xi\xi}, \quad \text{for} \quad z = h_0,\\
&\psi_{3z} + h_1\psi_{2zz} = h_{1\tau} - c_0 h_{2\xi} + \psi_{1\xi}h_{1\xi}, \quad \text{for} \quad z = h_0,\\
&-c_0\psi_{1\xi} + gh_1 = 0, \ \ \psi_{1\tau} - c_0\psi_{2\xi} + \psi_{1\xi}^2/2 + gh_2 = 0.
\end{aligned} \tag{3.4.11}$$

Elimination results in

$$\frac{\partial h_1}{\partial\tau} + \frac{3c_0}{2h_0}h_1\frac{\partial h_1}{\partial\xi} + \frac{c_0 h_0^2}{6}\frac{\partial^3 h_1}{\partial\xi^3} = 0. \tag{3.4.12}$$

We now use a simplified form

$$v_t + \alpha v v_x + v_{xxx} = 0. \tag{3.4.13}$$

This is the KORTEWEG-DE VRIES *equation* in its most usual form ($\alpha = -6$). This nonlinear partial differential equation of third order can be integrated using the setup

$$v(x,t) = v(\eta), \quad \eta = x - c_0 t. \tag{3.4.14}$$

This corresponds to a transformation to a co-moving wave frame. One obtains

$$v_\eta(\alpha v - c_0) + v_{\eta\eta\eta} = 0. \tag{3.4.15}$$

Double integration with respect to $\eta$ gives

$$v_{\eta\eta} = c_0 v - \alpha\frac{v^2}{2}, \quad \frac{1}{2}v_\eta^2 = \frac{c_0}{2}v^2 - \frac{\alpha}{6}v^3, \tag{3.4.16}$$

and finally

$$\int_{v_{\max}}^{v} \frac{dv}{\sqrt{c_0 v^2/2 - \alpha v^3/6}} = \sqrt{2}\,\eta. \tag{3.4.17}$$

This results in a *soliton* solution

$$v(x,t) = \frac{3c_0}{\alpha}\mathrm{sech}^2\left(\frac{\sqrt{c_0}}{2}(x - c_0 t)\right), \tag{3.4.18}$$

see Fig. 1.5 and Eq. (1.1.17). Other soliton solutions of (3.4.13) may be found in the literature [3.9]

$$v(x,t) = \frac{-2\alpha^2 f(x,t)}{(1 + f(x,t))^2}, \tag{3.4.19}$$

where

$$f(x,t) = \exp(-\alpha(x - x_1) + \alpha^3 t). \tag{3.4.20}$$

Here $x_1$ is an integration constant. The maximum absolute value of $v(x,t)$, its amplitude, is given by $\alpha^2/2$ and occurs when $f = 1$ or $x = \alpha^2 t + x_1$, where $x_1$ is now a phase shift. The propagation speed of the soliton (3.4.19) is $\alpha^2$ and thus proportional to its amplitude. The parameters $\alpha$ and $x_1$ are determined by the initial conditions.

If the initial condition is in the form of the soliton, i.e.,

$$v(x, t = 0) = v_0\mathrm{sech}^2\left(\frac{x - x_0}{\Delta}\right), \tag{3.4.21}$$

where $\Delta = 2\sqrt{\alpha/c}$ is the width of the wave, then this initial perturbation propagates unchanged like a soliton if its amplitude were connected with the velocity by

$$v_0\Delta^2 = 12\alpha. \tag{3.4.22}$$

Thus the larger the soliton amplitude $v_0$, the larger its velocity $c > c_0$, while a wave packet which spreads in time and decreases in amplitude is slower.

If the initial condition does not coincide in profile with the soliton, but has the form of a pulse of width and amplitude (earthquake, underwater explosion), then the so called *nonlinearity parameter* $\sigma$, a dimensionless quantity [3.10]

$$\sigma = \sqrt{\frac{v_0}{c_0}}k_0\Delta = \Delta\sqrt{\frac{v_0}{\alpha}} \tag{3.4.23}$$

is $\sigma > \sqrt{12} = \sqrt{2/\alpha}$. This is the value for a soliton. For $\sigma \ll \sqrt{12}$ one deals again with an almost linear perturbation. Great interest is however attached to the case of a strongly nonlinear perturbation characterized by $\sigma \gg \sqrt{12}$, when the width is large. During the first steps of the evolution of such a strong perturbation, dispersion does not yet play an important role and the development of the perturbation is determined by the nonlinearity: steepening occurs and the front has a tendency to break. Later on dispersion comes into play and the perturbation may break up into individual waves. With advancing time these individual waves become transformed into solitons. If dissipation is relevant, the picture may change.

In order to investigate the effects of dissipation due to viscosity on solitons, we first remember the BURGERS *equation* (3.2.50). We then add the dissipative term to the KORTEWEG-DE VRIES BURGERS *equation* [3.10]

$$u_t + uu_x + \beta u_{xxx} = \nu u_{xx}. \tag{3.4.24}$$

This equation describes a dissipative soliton.

In the literature solitons are very often described by the BOUSSINESQ *equations.* There exist many different forms of these equations which were derived by BOUSSINESQ in the year 1870 to model the propagation of long water waves with a small amplitude (– sign) and to describe the two-dimensional irrotational flow of an inviscid liquid in a uniform rectangular channel (+ sign). The original equations are [3.21]

$$u_{tt} - u_{xx} \pm u_{xxxx} - (u^2)_{xx} = 0. \tag{3.4.25}$$

An improved modified version reads [3.22]

$$u_{tt} - u_{xx} - u_{xxtt} = (u^k)_{xx}, \quad k = 2 \text{ or } 3. \tag{3.4.26}$$

The *linear* BOUSSINESQ *equation* is defined by mathworld [3.21] as

$$u_{tt} - \alpha^2 u_{xx} = \beta^2 u_{xxtt}, \tag{3.4.27}$$

whereas the nonlinear one is assumed to be

$$u_{tt} - u_{xx} - u_{xxxx} - 3(u^2)_{xx} = 0. \tag{3.4.28}$$

There is also a modified equation reading

$$\frac{1}{3}u_{tt} - u_t u_{xx} - \frac{3}{2}u_x^2 u_{xx} + u_{xxxx} = 0. \tag{3.4.29}$$

Wolfram research [3.21] defines the BOUSSINESQ *approximation* that density variations are ignored, except insofar as they give rise to a gravitational force. This is also well known in plasma physics [2.8], [3.22]. *Classical* BOUSSINESQ *equations* are given by DIAS [3.13]

$$u_t + uu_x + g\eta_x - \frac{1}{3}h^2 u_{xxt} = 0, \quad (3.4.30)$$

$$\eta_t + [u(h+\eta)]_x = 0, \quad (3.4.31)$$

where $u$ had been replaced with the depth averaged velocity $(1/h)\int_{-h}^{\eta} u dz$, $h$ is the depth and $z = \eta$ designates the water surface.

Similar equations may be found in [3.20]

$$\frac{\partial \vec{v}}{\partial t} + (\vec{v}\,\nabla)\vec{v} + \nabla gH + \frac{gh^2}{3}\Delta\nabla H = 0, \quad (3.4.32)$$

$$\frac{\partial H}{\partial t} + \nabla(H\vec{v}) = 0. \quad (3.4.33)$$

Here $h$ is the depth of the fluid and $H$ is the total height of the fluid above bottom. Another form is [3.20]

$$\frac{\partial \rho}{\partial t} + \nabla(\rho\vec{v}) = 0, \quad (3.4.34)$$

$$\frac{\partial \vec{v}}{\partial t} + (\vec{v}\,\nabla)\vec{v} + \frac{c^2(\rho)}{\rho}\nabla\rho + \frac{2c_0\beta}{\rho_0}\nabla\Delta\rho = 0, \quad (3.4.35)$$

where $\rho$ is the fluid density, $c(\rho)$ is the generalized sound velocity at the density $\rho$, $\rho_0$ is the unperturbed density, $\beta$ is the dispersion parameter. Following KARPMAN [3.20] one can show, that the dispersion belonging to (3.4.35) is given by

$$\omega = c_0 k - \beta k^3. \quad (3.4.36)$$

Using the adiabatic law

$$c^2(\rho) = c_0^2 \cdot \left(\frac{\rho}{\rho_o}\right)^{\kappa-1}. \quad (3.4.37)$$

KARPMAN derives the following *potential equation*

$$\varphi_{tt} - c_0^2\Delta\varphi + \nabla\varphi\cdot\nabla\varphi_t + \kappa\Delta\varphi\cdot\varphi_t - 2c_0\beta\Delta^2\varphi = 0 \quad (3.4.38)$$

for $\varphi$ defined by $\vec{v} = \nabla\varphi$. Assuming a *travelling wave*

$$\varphi = \varphi(x - Vt) \quad (3.4.39)$$

the potential equation delivers a stationary solution

$$\left(v^2 - c_0^2\right)\varphi'' - (\kappa+1)V\varphi''\varphi' - 2c_0\beta\varphi'''' = 0. \quad (3.4.40)$$

If the wave amplitude is small, the propagation speed $V$ would be close to $c_0$, so that $V - c_0$ can be regarded as being small. Introducing a new variable

$$u = \frac{\kappa + 1}{2} \varphi' \tag{3.4.41}$$

one obtains

$$\beta u''' + uu' - (V - c_0)u' = 0. \tag{3.4.42}$$

This equation is invariant with respect to a translation

$$u \to u + \text{const}, \quad V \to V + \text{const}. \tag{3.4.43}$$

This signifies a transition to a new coordinate system. Integrating (3.4.42) twice one obtains

$$3\beta u'^2 = (b_1 - u)(b_2 - u)(b_3 - u), \tag{3.4.44}$$

where the three $b_i$ are constants consisting of the propagation speed $V$ and two integration constants. One has

$$V = \frac{1}{2}(b_1 + b_2 + b_3) + c_0. \tag{3.4.45}$$

Now several cases will be discussed.

1. We assume $\beta > 0$, $b_1 \geq b_2 \geq b_3$, $b_2 \leq u \leq b_1$ and $b_2 = b_3$. Then the solution of (3.4.44) is a *soliton*

$$u(\xi) = \frac{b_1 - b_2}{\cosh^2\left(\sqrt{(b_1 - b_2)/12\beta}\xi\right)} + b_3. \tag{3.4.46}$$

   The form of this solution is shown in Fig. 1.5, since ch = cosh = csch = 1/sech [1.4] [2.1].

2. When $b_3 < b_2$, the solution of (3.4.44) is periodic and expressed by

$$u(\xi) = \frac{b_1 - b_2}{s^2} \mathrm{dn}^2(z; s) + b_3. \tag{3.4.47}$$

   Here $s$ is the modulus and dn is the *elliptic* JACOBI *function* . Furthermore

$$s^2 = \frac{b_1 - b_2}{b_1 - b_3}, \quad z = \sqrt{\frac{b_1 - b_2}{12\beta}} \frac{\xi}{s}. \tag{3.4.48}$$

The function $\mathrm{dn}^2$ is periodic with a period $2K(s)$, where $K$ is the complete *elliptic integral* of first order [1.1], [1.4]. The wave length $\lambda$ of (3.4.47) is then

$$\lambda = 2\sqrt{\frac{12\beta}{(b_1 - b_2)}} sK(s). \tag{3.4.49}$$

The Mathematica commands

```
U[x_]=(JacobiDN[x,s])^2;
Clear[s],s=0.9;
Plot[U[x],{x,-13.,13.}]                                    (3.4.50)
```

produce Fig. 3.3, which shows the periodic solution (3.4.47). Compare with Fig. 1.6.

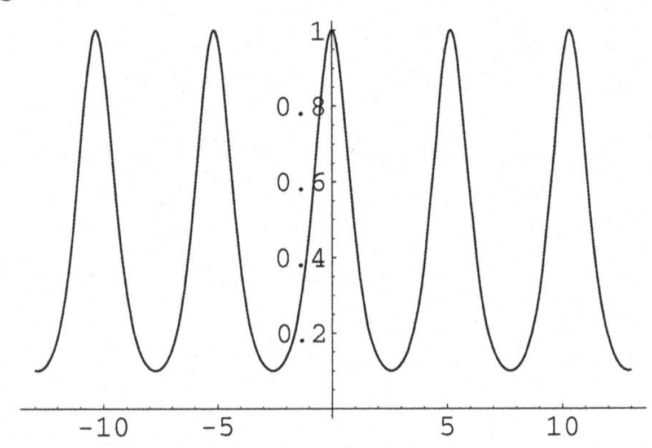


Fig. 3.3. Periodic soliton

3. For $\beta < 0$, and the substitutions $\xi \to -\xi$, $u \to -u$, $V - c_0 \to -(V - c_0)$ one obtains a soliton showing the same form as in Fig. 1.5.

The *phase portrait* $u'(u)$ [1.1] of the differential equation (3.4.44) exhibits the special character of the soliton. Such a portrait can be created by the Mathematica commands

```
<<Graphics'ImplicitPlot'
Clear[b1,b2,b3,β,p1]
b1=7.0;b2=1.5;b3=-1.8;β=11;
F[u_,v_]=3*β*v^2-(b1-u)*(b2-u)*(b3-u);
p1=ImplicitPlot[F[u,v]==0, {u,-5.,10.}]
<<Graphics'ImplicitPlot'
Clear[b1,b2,b3,β,p2]
b1=9.0;b2=0.;b3=0.;β=3;
```

```
F[u_,v_]=3*β*v^2-(b1-u)*(b2-u)*(b3-u);
p2=ImplicitPlot[F[u,v]==0, {u,-5.,10.}]
<<Graphics'ImplicitPlot'
Clear[b1,b2,b3,β,p3]
b1=6.;b2=3.;b3=-0.8;β=6.5;
F[u_,v_]=3*β*v^2-(b1-u)*(b2-u)*(b3-u);
p3=ImplicitPlot[F[u,v]==0, {u,-5.,10.}]
Show[p1,p2,p3]
```

The last command combines the three plots p1, p2, p3 and creates Fig. 3.4.

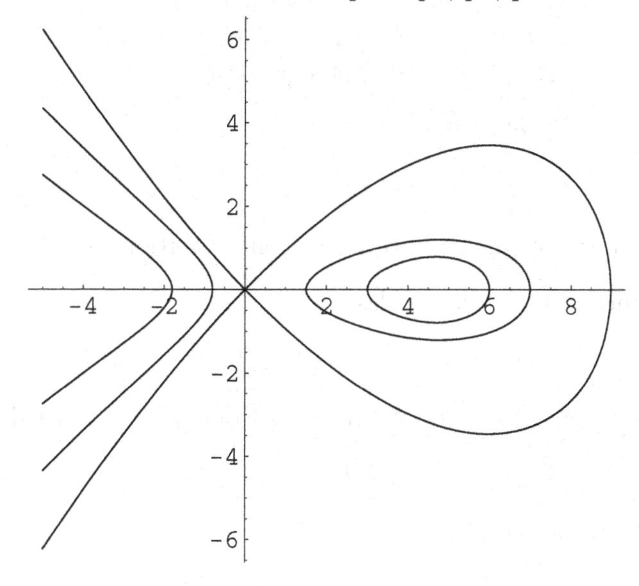


Fig. 3.4. Phase portrait of (3.4.44)

The soliton portrait is given by the curve with the branching point (cross-over point), which is called a *saddle point* in a portrait ([1.1, p. 72]). The closed curves in the phase portrait describing periodic waves are connected with *limit cycles* and the BENDIXON *criterion* [1.1].

## Problems

1. Transform the partial differential equation (3.4.24) into an ordinary differential equation using the setup for a *travelling wave*

$$u(x-ct) = u(\zeta). \tag{3.4.51}$$

Solution: after integration with respect to $\zeta$

$$-cu + u^2/2 + \beta u'' = \nu u'. \tag{3.4.52}$$

This equation describes a damped nonlinear oscillator.

2. Is it possible to solve (3.4.24) by a *similarity transformation*?

$$u(x,t) = x^\alpha t^\gamma. \tag{3.4.53}$$

Solution: no. Why?

3. Verify (3.4.7) to (3.4.9). Hint: use $\varphi(x,z,t) = \psi(\xi(x,t),z,\tau(t))/\sqrt{\varepsilon}$, $h(x,z,t) = h(\xi(x,t),z,\tau(t))$ and calculate $\partial\tau/\partial t$, $\partial\xi/\partial x$, $\partial\xi/\partial\tau$ and insert $\varphi_x, \varphi_z, h_x, h_t$ etc into the equations.

4. Introduce *stretched variables*

$$\xi = \varepsilon^n(x - c\tau), \quad \tau = \varepsilon^m t, \quad U = \varepsilon^p u. \tag{3.4.54}$$

Which term of (3.4.24) prevents such a solution?

5. Introduce *stretched variables*

$$\eta = u/u_0, \quad \xi = x/l, \quad \tau = tu_0/l \tag{3.4.55}$$

into the KORTEWEG-DE VRIES equation (3.4.13) to read

$$u_t + hu_x + \beta u_{xxx} = 0 \tag{3.4.56}$$

and the initial condition

$$u(x,0) = u_0\varphi(x/l). \tag{3.4.57}$$

Solution:

$$\eta_t + \eta\eta_\xi + \frac{1}{\sigma^2}\eta_{\xi\xi\xi} = 0, \quad \eta(\xi,0) = \varphi(\xi). \tag{3.4.58}$$

The dimensionless parameter

$$\sigma = l\sqrt{u_0/\beta} \tag{3.4.59}$$

is now the *nonlinearity parameter*, compare with (3.4.23). This parameter determines the relation between nonlinearity and dispersion. For the balancing of the two effects (no change of form of the soliton) the condition $\sigma = \sigma_s = \sqrt{12}$ must be satisfied. But this value depends on the form of the KORTEWEG-DE VRIES equation.

6. Transform (3.4.56) using *stretched variables* and a *similarity transformation*

$$t \to \gamma t, \quad x \to \gamma^{1/3} x, \quad u \to \gamma^{-2/3} u, \quad z = (3\beta t)^{-1/3} x \tag{3.4.60}$$

and

$$u(x,t) = \beta(3\beta t)^{-2/3} \psi[(3\beta t)^{-1/3} x]. \tag{3.4.61}$$

Solution:

$$\psi''' - z\psi' + \psi\psi' - 2\psi = 0. \tag{3.4.62}$$

7. Neglect the nonlinear term in (3.4.62) and insert $\psi(z) = f'(z)$ into (3.4.62).

Solution:

$$f''(z) - zf(z) = 0 \tag{3.4.63}$$

which is the equation determining AIRY *functions*. Solve (3.4.63).

Solution:

$$f(z) = Ai(z), \tag{3.4.64}$$

see [1.4]. Plot $f(z)$ using Mathematica.

8. The KORTEWEG-DE VRIES BURGERS *equation* (3.4.24) may be regarded as the dissipative KORTEWEG-DE VRIES equation. What is the effect of the damping term due to viscosity? Make a setup for a *travelling wave* solution

$$u = u(x - Wt). \tag{3.4.65}$$

Solution:

$$\beta u''' - \nu u'' + u'(u - W) = 0. \tag{3.4.66}$$

Assume $u(\infty) = u'(\infty) = u''(\infty) = 0$ and integrate.

Solution:

$$\beta u'' - \nu u' + \frac{1}{2}u^2 - Wu = 0. \tag{3.4.67}$$

Now the structure of the solution $u$ depends on the relation between the dispersion parameter $\beta$ and the dissipative parameter $\nu$ (viscosity). The wave velocity $W$ is not constant but depends on $u$.

9. An analytic solution of the KORTEWEG-DE VRIES equation has been found by the *inverse scattering method.* This method is an ingenious method to solve nonlinear partial differential equations. The method does not directly solve the nonlinear equation, but instead, it solves two linear equations. It considers the assumed solution $v$ of $v_t - 6vv_x + v_{xxx} = 0$ to be the potential $v(x,t)$ in a one-dimensional SCHROEDINGER *equation* $\psi_{xx} + [E - v(x,t)]\psi = 0$. Study the relevant pages, e.g. in[1.1], [3.20].

## *3.5 Dissipationless tsunamis*

A *tsunami* ("large wave in the harbour") is a large amplitude (nonlinear) gravitational wave excited by an earthquake or an underwater explosion. Its local crest height and propagation speed depend strongly on the ocean depth. The crest may vary between less than one m up to 30 m near cost and the speed between several hundred to 1000 km h$^{-1}$. As long as no dissipative effects are included into the mathematical analysis it can be assumed that the thermodynamic behavior is polytropic - entropy is constant and a velocity potential exists. For water the polytropic exponent is equal to two.

Tsunamis are not of the type of surf swells – they move the entire depth of the ocean down to several km depth. There may be several hours time passing between the creation and its impact on a far distant coast. The seismic wave may arrive there faster than the tsunami itself. In open water the wave period of a tsunami may range from minutes to hours and the wavelengths may be up to several hundred km.

In order to derive equations for a one-dimensional tsunami wave in the $x$-direction we define $\eta(x,t)$ as the local distance between the horizontal $x$-axis and the water surface, see Fig. 3.5.

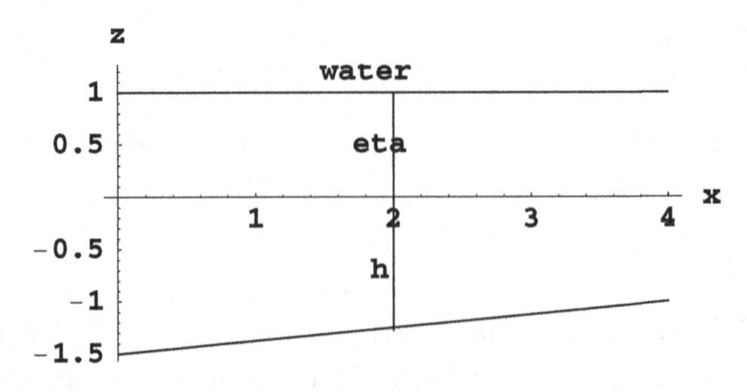


Fig. 3.5. Water surface (eta = $\eta$)

This figure has been generated by the following Mathematica commands:

```
Clear[l1,l2,l3,l4,T1,T2,T3,G1,G2,G3,G4,G5,G6,G7];
l1=Line[{{0.,1.},{4.,1.}}];
l2=Line[{{0.,-1.5},{4.,-1.}}];
l3=Line[{{2.,0.},{2.,1.}}];
l4=Line[{{2.,-1.28},{2.,0.}}];
T1=Text[water,{1.8,1.2}];
T2=Text[eta,{1.9,-0.5}];
T3=Text[h{1.9,-0.7}];
$DefaultFont={''Courier-Bold'',10};
G1=Graphics[T1];
G2=Graphics[T2];
G3=Graphics[T3];
G4=Graphics[l1,Axes->True,AxesLabel->{''x'',''z''},
AspectRatio->0.5];
G5=Graphics[l2,Axes->True,AxesLabel->{''x'',''z''},
AspectRatio->0.5];
G6=Graphics[l3,Axes->True,AxesLabel->{''x'',''z''},
AspectRatio->0.5];
G7=Graphics[l4,Axes->True,AxesLabel->{''x'',''z''},
AspectRatio->0.5];
Show[G4,G5,G6,G7,G1,G2,G3]
```

The local distance between the variable ocean bottom and the $x$-axis is designed by $-h(x)$. This is a given quantity and describes a boundary condition. Now the relation

$$z = -h(x, y) \tag{3.5.1}$$

holds and the local hydrostastic pressure is given by

$$p(x, z, t) = \rho_0 g(\eta + h) = \rho_0 g(\eta - z), \tag{3.5.2}$$

since gravity is assumed to act in the $z$-direction. On the water surface $z = \eta$ the hydrostatic pressure vanishes.

For a *dissipationless tsunami* one has $\operatorname{curl} \vec{v} = 0$, so that (2.3.18) gives the equation of motion in the form [3.11]

$$\frac{\mathrm{d}\vec{v}}{\mathrm{d}t} = \frac{\partial \vec{v}}{\partial t} + \nabla \frac{\vec{v}^2}{2} = -\frac{\nabla p}{\rho_0} = -g\nabla\eta. \tag{3.5.3}$$

The CORIOLIS *force* has been neglected.

If we assume a two-dimensional flow we write $v_x = u$, $v_z = w$. Then (3.5.3) yields the components

$$\frac{\mathrm{d}u}{\mathrm{d}t} = \frac{\partial u}{\partial t} + \frac{1}{2}\frac{\partial}{\partial x}(u^2 + w^2) = -g\frac{\partial \eta}{\partial x} \tag{3.5.4}$$

and

$$\frac{\mathrm{d}w}{\mathrm{d}t} = \frac{\partial w}{\partial t} + \frac{1}{2}\frac{\partial}{\partial z}(u^2 + w^2) = -\rho_0 g - \frac{\partial p}{\partial z}. \tag{3.5.5}$$

Here $h(x, y)$ is given and $u(x, z, t)$, $w(x, z, t)$ and $\eta(x, z, t)$ are the unknowns. The third equation is given by the continuity equation

$$\nabla \vec{v} = \operatorname{div} \vec{v} = 0. \tag{3.5.6}$$

If the vertical water particle acceleration could be neglected (if $p_x$ is independent of $z$), then (3.5.5) breaks down and $u$ is independent of $z$. Then for shallow water (3.5.4) reads ($w^2 \approx 0$)

$$u_t + uu_x = -g\eta_x \tag{3.5.7}$$

and (3.5.6) takes the form

$$u_x = -w_z. \tag{3.5.8}$$

Integration with respect to $z$ gives [3.11]

$$w(\eta) - w(-h) = -\int_{-h}^{\eta} u_x \mathrm{d}z = -u_x(\eta + h). \tag{3.5.9}$$

Now we consider the boundary conditions. On the ocean bottom $z = -h(x)$, the normal component of the fluid velocity vanishes. Due to the inclination of the bottom, the simple condition $w = 0$ for $z = -h(x)$ is not valid. The angle $\vartheta$ of the inclination is given by (3.1.1). Thus

$$\tan \vartheta = \frac{\partial h(x)}{\partial x} = -\frac{w}{u} \tag{3.5.10}$$

and the *bottom boundary condition* reads

$$h_x u + w = 0, \quad \text{for} \quad z = -h(x). \tag{3.5.11}$$

The boundary condition on the free water surface is given by

$$z = \eta(x, t). \tag{3.5.12}$$

This expresses the fact that a water particle residing on the surface will remain there for all time (mass conservation). Thus for $z - \eta = 0$ one has $\mathrm{d}/\mathrm{d}t(z - \eta) = 0$. Since $\eta_z = 0$, one obtains for the *surface condition*

$$\eta_t + u\eta_x - w = 0, \quad \text{for} \quad z = \eta. \tag{3.5.13}$$

Sometimes this condition is called *kinematic condition.* $p = 0$ at $z = \eta$ is called *dynamic condition* [3.13]. Using the BERNOULLI equation on the free surface $z = \eta$ gives $\varphi_t + (1/2)(\nabla\rho)^2 + g\eta = 0$ [3.13]. Using the two boundary conditions (3.5.11) and (3.5.13) in the form

$$w(h) = -h_x u \tag{3.5.14}$$

and

$$w(\eta) = \eta_t + u\eta_x \tag{3.5.15}$$

we can insert two expressions into the continuity equation (3.5.9) to obtain

$$\eta_t + u\eta_x + h_x u = -u_x(\eta + h),$$
$$\eta_t + \frac{\partial}{\partial x}((\eta + h)u) = 0. \tag{3.5.16}$$

Based on the equivalence theorem of section 2.10, the equations (3.5.7) and (3.5.16) may be rewritten in the form

$$\hat{\rho}(u_t + uu_x) = -\hat{p}_x, \tag{3.5.17}$$

$$\hat{\rho}_t + \frac{\partial}{\partial x}(\hat{\rho}u) = 0. \tag{3.5.18}$$

For $w \approx 0$ the equations (3.5.7) and (3.5.16) describe *roll waves* defined by (3.1.3) and (3.1.4) and do not describe a tsunami.

To describe a tsunami the following items must be taken into account:

1. vertical motion of the water, is present, $w \neq 0$,
2. the ocean bottom is neither a horizontal nor a declined plane, but it has a topographic structure $h(x, y)$,
3. the path of a tsunami over the surface of the rotating Earth is co-determined by the CORIOLIS- and the centrifugal force,
4. dissipation due to the water viscosity and, if necessary, the thermal conductivity has to be considered, so that real life tsunamis are no longer polytropic,
5. wave steepening and modifications of wavelength (dispersion) etc have to be considered,

6. the theory must be able to reproduce observed tsunami data like

| | |
|---|---|
| wave length | 1 - 40 - 160 - 240 km |
| ocean depth | 0.1 - 4 km |
| coastal depth | 5 cm - 5 m |
| crest height | 0.5 m in open ocean |
| | up to 30 m on the coast (depth 0.5 m) |
| propagation speed | 365 $\mathrm{km\,h^{-1}}$ at depth 1 km |
| | 750 - 900 $\mathrm{km\,h^{-1}}$ at depth 4 - 5 km |

For $h = 100$, $\eta = 10$ one obtains from $c = \sqrt{g(h+\eta)}$ 118 $\mathrm{km\,h^{-1}}$ and for $h = 50$, $\eta = 30$ one obtains 100 $\mathrm{km\,h^{-1}}$. One has, however, to have in mind, that the observed wave propagation speed $w$ is actually defined by $w = u + c$.

These observed values coincide with the results given by (3.2.22). For $h = 4000\,\mathrm{m}$ one obtains 713 $\mathrm{km\,h^{-1}}$ and for $h = 1000\,\mathrm{m}$ one obtains 357 $\mathrm{km\,h^{-1}}$. This seems to prove that the neglection of the vertical velocity component $w$ does not entail a large error.

In order to start a deeper investigation of the tsunami problem we now first undertake to derive a *potential equation for tsunamis.* We use the following equations which made use of the equivalence principle. We follow the method used in [1.1]. Since the bottom $h$ now depends on $x$ and $y$, (3.5.2) becomes

$$p(x,y,z,t) = \rho_0 g(\eta(x,y,t) - z) \tag{3.5.19}$$

and $\eta(x,y,t) = z - h(x,y)$. Furthermore the continuity equation now reads

$$\frac{1}{\eta}\eta_t - \Delta\varphi - \nabla\varphi\frac{\nabla\eta}{\eta} = 0, \tag{3.5.20}$$

compare with (2.8.1). In the derivation we have used

$$\vec{v} = -\nabla\varphi, \qquad \varphi = \varphi(x,y,z,t) \tag{3.5.21}$$

and the replacement $\rho \to \hat{\rho} \to \rho_0\eta(x,y,t)$, $\operatorname{div}\operatorname{grad} = \Delta$. The equation of motion reads now

$$-\nabla\frac{\partial\varphi}{\partial t} + \frac{1}{2}\nabla(\nabla\varphi)^2 = -g\nabla\eta \tag{3.5.22}$$

because $\nabla z = 0$ and $(\vec{v}\,\nabla)\vec{v} = (1/2)(\nabla\varphi)^2$ for a potential flow, compare with (2.10.29)! Integration with respect to space yields an equation which may also be called time dependent BERNOULLI *equation*

$$-\frac{\partial\varphi}{\partial t} + \frac{1}{2}(\nabla\varphi)^2 = -g\eta, \tag{3.5.23}$$

compare with (2.8.4). See also [3.13]. Derivation with respect to time gives

$$-\frac{\partial^2\varphi}{\partial t^2}+\frac{1}{2}\frac{\partial}{\partial t}(\nabla\varphi)^2=-g\frac{\partial\eta}{\partial t}, \tag{3.5.24}$$

compare with (2.8.5). Now we build

$$\frac{1}{\eta}\eta_t=\frac{1}{g\eta}\varphi_{tt}-\frac{1}{g\eta}\frac{1}{2}\frac{\partial}{\partial t}(\nabla\varphi)^2. \tag{3.5.25}$$

From (3.5.24) and (3.5.22) we receive

$$\frac{\nabla\eta}{\eta}=\frac{1}{g\eta}\left(\nabla\varphi_t-\frac{1}{2}\nabla(\nabla\varphi)^2\right). \tag{3.5.26}$$

Some remarks on units might be appropriate: pressure $p$ has $\mathrm{kg\,s^{-2}\,m\,s^{-1}}$, density $\rho_0$ has $\mathrm{kg\,m^{-3}}$, $\eta$ is measured in m and $u$ in $\mathrm{m\,s^{-1}}$. Thus the potential $\varphi$ has $\mathrm{m^2\,s^{-1}}$.

Inserting from (3.5.25) and (3.5.36) into (3.5.20) we obtain the *potential equation*

$$-\frac{1}{c^2}\frac{1}{2}\frac{\partial}{\partial t}(\nabla\varphi)^2-\frac{1}{c^2}\nabla\varphi\cdot\nabla\varphi_t+\frac{1}{c^2}\nabla\varphi\cdot\frac{1}{2}\nabla(\nabla\varphi)^2=\frac{1}{c^2}\varphi_{tt}-\Delta\varphi, \tag{3.5.27}$$

compare (2.8.9). Here we used $h\to\eta$ and(3.5.29)

$$c^2=g\eta(x,y,t), \tag{3.5.28}$$

compare with (3.2.22). In components (3.5.27) reads

$$\begin{aligned}&\varphi_{xx}(c^2-\varphi_x^2)+\varphi_{yy}(c^2-\varphi_y^2)+\varphi_{zz}(c^2-\varphi_z^2)-2\varphi_x\varphi_y\varphi_{xy}\\&-2\varphi_y\varphi_z\varphi_{yz}-2\varphi_x\varphi_z\varphi_{xz}-\varphi_{tt}-2\varphi_x\varphi_{xt}-2\varphi_y\varphi_{yt}-2\varphi_z\varphi_{zt}=0.\end{aligned} \tag{3.5.29}$$

For only the two independent variables $x, y$ this equation becomes identical with the results by PRESSWERK of January 1938 [2.21]. It is a drawback of equation (2.3.29) that it cannot be linearized by a LEGENDRE *transformation* because this works only for two independent variables $x, y$ or $x, t$ like in [2.21].

The boundary conditions for the potential $\varphi(x, y, z, t)$ for the bottom $z=-h(x, y)$ are $\nabla\varphi\cdot\nabla h=-\varphi_z$ or

$$\frac{\partial h(x,y)}{\partial x}=-\frac{w}{u}=-\frac{\varphi_z}{\varphi_x}, \tag{3.5.30}$$

$$\frac{\partial h(x,y)}{\partial y}=-\frac{w}{v}=-\frac{\varphi_z}{\varphi_y}, \tag{3.5.31}$$

where the components of the vector $\vec{v}$ have been designated by $u, v, w$. On the water surface $z = \eta$ one has a free (moving) boundary. This condition may be expressed in the three-dimensional form

$$\eta_t + \nabla\varphi \cdot \nabla\eta - \varphi_z = 0, \tag{3.5.32}$$

compare with (3.5.13) [3.13] and with the BERNOULLI equation (3.5.23) as well as for $z = \eta$. Some authors drop square terms and use linearized free-surface conditions [3.14]. Additionally one has initial conditions at $t = 0$ for $\varphi$ (and $\eta$) needed to handle the tsunami problem.

Since more than 40 years numerical methods are available for the numerical solution of systems of quasilinear hyperbolic partial differential equations [3.12], [3.15], [3.18]. The method of characteristics for two independent variables which we used in previous sections cannot be generalized because the theory for characteristics in two variables is a very special case. In higher dimensions new types of characteristics like bicharacteristics and characteristic cones appear. Since my small University does not have the necessary high power computer facilities, we are looking for other possibilities to handle the tsunami problem.

In order to derive a simple *tsunami wave equation*, we consider the case of two independent variables $x, t$. Having confidence in the equivalence principle we consider the one-dimensional time dependent flow of a compressible gas [2.18] described by $\rho(x,t)$ and $u(x,t)$. The continuity equation (2.8.1) takes the form

$$\frac{\mathrm{d}\rho}{\mathrm{d}t} + \rho u_x = 0, \tag{3.5.33}$$

and the equation of motion (2.8.2) is

$$\frac{\mathrm{d}u}{\mathrm{d}t} + \frac{1}{\rho} p_x = 0. \tag{3.5.34}$$

These two quasilinear partial differential equations of second order may be transformed into a nonlinear partial differential equation of second order with two independent variables ("wave equation"). The operator

$$\frac{\mathrm{d}}{\mathrm{d}t} = \frac{\partial}{\partial t} + u\frac{\partial}{\partial x} \tag{3.5.35}$$

contains the nonlinearity. To transform away the nonlinearity we define a *mass variable* $m$ by

$$m(x,t) = \int_{x_0}^{x} \rho(x,t)\mathrm{d}x. \tag{3.5.36}$$

Here we assumed that the volume which contains $m$ has an extension of 1 cm in the $y$- and $z$-direction. Due to the continuity equation the mass contained in this volume for $x_0$ to $x$ is constant. Thus $\mathrm{d}m/\mathrm{d}t = 0$. Then

$$\frac{\mathrm{d}m(x,t)}{\mathrm{d}t} = \frac{\partial m}{\partial x}\frac{\mathrm{d}x}{\mathrm{d}t} + \frac{\partial m}{\partial t} = 0. \tag{3.5.37}$$

Now (3.5.36) has the consequence

$$\left(\frac{\partial m}{\partial x}\right)_t = \rho(x,t). \tag{3.5.38}$$

Then (3.5.37) yields

$$\left(\frac{\partial m}{\partial t}\right)_x = -\left(\frac{\partial m}{\partial x}\right)_t \frac{\mathrm{d}x}{\mathrm{d}t} = -\rho u. \tag{3.5.39}$$

As we will demonstrate, the transformation of the two independent variables $x, t$ into new variables $m, t$ will allow the important transition $(\mathrm{d}/\mathrm{d}t) \to (\partial/\partial t)$ and $u(x,t) \to \hat{u}(m(x,t),t)$ as well as $\rho(x,t) \to \hat{\rho}(m(x,t),t)$. We thus have

$$u_x = \hat{u}_m \cdot m_x = \hat{u}_m \hat{\rho}, \quad \rho_x = \hat{\rho}_m \cdot \hat{\rho}, \quad p_x = \hat{p}_m \cdot \hat{\rho}. \tag{3.5.40}$$

The more interesting result is however given by

$$\begin{aligned} u_t &= \hat{u}_m \cdot m_t + \hat{u}_t = -\hat{u}_m \cdot \hat{\rho}\hat{u} + \hat{u}_t, \\ \rho_t &= \hat{\rho}_m \cdot m_t + \hat{\rho}_t = -\hat{\rho}_m \cdot \hat{\rho}\hat{u} + \hat{\rho}_t. \end{aligned} \tag{3.5.41}$$

In fact, one has

$$\frac{\mathrm{d}u}{\mathrm{d}t} = \frac{\partial u}{\partial t} + u\frac{\partial u}{\partial x} = -\hat{u}_m\hat{\rho}\hat{u} + \hat{u}_t + \hat{u}\hat{u}_m\hat{\rho} = u_t,$$

so that formally

$$\frac{\mathrm{d}u}{\mathrm{d}t} = \frac{\partial \hat{u}}{\partial t} \quad \text{and} \quad \frac{\partial}{\partial x} = \hat{\rho}\frac{\partial}{\partial m}$$

are valid. Applying (3.5.35) on (3.5.33) and then inserting $\rho_t = \hat{\rho}_t, \hat{u}, \hat{\rho}_x$ and $\hat{u}_x$ one obtains the continuity equation in the form

$$\hat{\rho}_t + \hat{\rho}^2\hat{u}_m = 0. \tag{3.5.42}$$

The equation of motion (3.5.34) now reads

$$\hat{u}_t + \hat{p}_m = 0 \tag{3.5.43}$$

after transformation. Here we had used

$$\frac{\partial p(x,t)}{\partial x} = \frac{\partial \hat{p}(m(x,t),t)}{\partial m} \cdot \frac{\partial m}{\partial x} = \hat{p}_m \hat{\rho}. \tag{3.5.44}$$

Equation (3.5.43) is a linear partial differential equation, but (3.5.42) is not! But now we use the transformation (2.2.7) to linearize (3.5.42). Since $s(x,t) = 1/(\rho(x,t)$ and

$$\rho(x,t) = \hat{\rho}(m(x,t),t) = 1/\hat{s}(m(x,t),t), \tag{3.5.45}$$

and obtains from (3.5.42) the exactly linearized form of the continuity equation

$$\hat{u}_m - \hat{s}_t = 0. \tag{3.5.46}$$

Now the path to a wave equation for the compressible gas is wide open.

Method 1: We define a new function $W(m,t)$ by

$$W_t(m,t) = \hat{u}(m,t), \quad W_m(m,t) = \hat{s}(m,t), \tag{3.5.47}$$

which satisfies immediately (3.5.46) due to $W_{tm} = W_{mt}$. From (3.5.43) we obtain

$$W_{tt} = -\hat{p}_m. \tag{3.5.48}$$

Now we need the connection between $\hat{p}$ and $\hat{s}$ or $W_m$. For adiabatic change of state of a gas one obtains from (2.5.44)

$$\hat{p} = p = \frac{p_0}{\rho_0^\kappa}\rho^\kappa = p_0 s_0^\kappa s^{-\kappa} = \text{const} \cdot s^{-\kappa}. \tag{3.5.49}$$

Then (3.5.48) yields

$$\frac{\text{const}}{W_m^{-\kappa-1}} W_{tt} = W_{mm}. \tag{3.5.50}$$

Using the formal abbreviation

$$\frac{\text{const}}{W_m^{-\kappa-1}} = \frac{1}{a^2}, \tag{3.5.51}$$

then (3.5.50) takes the form of a second order wave equation

$$\frac{1}{a^2} W_{tt} = W_{mm}. \tag{3.5.52}$$

Method 2: One may define another $W(m,t)$ by

$$\hat{u} = W_{mt}, \quad \hat{s} = W_{mm}. \tag{3.5.53}$$

This satisfies (3.5.43) and (3.5.46). Using (3.5.46) one obtains after integration with respect to $m$ [2.18]

$$-W_{tt} = \text{const} \cdot W_{mm}^{-\kappa}, \quad \text{const} = a^2/\kappa. \tag{3.5.54}$$

It is obvious, that the simple assumption $\kappa = 2$ would not help to solve the tsunami problem, because the equation (3.5.54) has not yet been solved for $\kappa = 2$ and the tsunami needs a three-dimensional treatment. Equation (3.5.54) may be transformed into the DARBOUX equation which cannot be solved for $\kappa = 2$. It may be of interest to repeat the previous procedure with a mass variable for the tsunami equations, see problem 15 in this section!

Recently, an interesting set of equations describing tsunamis has been published by DIAS [3.13]. This three-dimensional formulation presents splitting up of the $z$-coordinate

$$\nabla = \frac{\partial}{\partial x}, \frac{\partial}{\partial y}, \quad \vec{u} = u, v. \tag{3.5.55}$$

The equation of continuity is given in the form

$$\nabla \cdot \vec{u} + \frac{\partial w}{\partial z} = 0 \tag{3.5.56}$$

and the equation of motion reads

$$\rho_0 \frac{\mathrm{d}\vec{u}}{\mathrm{d}t} = \rho_0 \frac{\partial \vec{u}}{\partial t} + \vec{u}(\nabla \vec{u}) + \nabla p = 0, \quad \rho_0 \frac{\mathrm{d}w}{\mathrm{d}t} + \rho_0 g + \frac{\partial p}{\partial z} = 0. \tag{3.5.57}$$

The water density $\rho_0$ is assumed to be constant. Since the author assumes a dissipationless tsunami, he uses a potential $\varphi(x, y, z, t)$

$$\vec{u} = \nabla \varphi, \quad w = \frac{\partial \varphi}{\partial z}. \tag{3.5.58}$$

The potential satisfies the LAPLACE *equation* in three dimensions $\Delta\varphi + \varphi_{zz} = 0$. The integration of the equation of motion gives BERNOULLI's *equation*

$$\varphi_t + \frac{1}{2}(\nabla\varphi)^2 + \frac{1}{2}\varphi_z^2 + gz + p/\rho_0 = 0. \tag{3.5.59}$$

Finally, the three boundary conditions have been used: the kinematic conditions (3.5.32), the dynamic condition (3.5.23), both at the water surface defined by $z = \eta(x, y, t)$. The bottom boundary condition at $z = -h(x, y)$ is given in the form

$$\nabla\varphi \cdot \nabla h + \varphi_z = 0, \tag{3.5.60}$$

compare with (3.5.30) and (3.5.31).

The solution of these equations is found by series expansions of the form $\varphi = \varphi_0 + \beta\varphi_1 + \beta^2\varphi_2$. An interesting point of this paper is, that a solution for $\varphi$ is found without knowing the function $p$. Only at the end of all calculations the equation of motion is again used to find the pressure from $\eta$ and $\varphi$. The equations (3.5.56) and (3.5.60) are the four *basic tsunami equations* describing $\vec{v}(x, y, z, t)$ and $\eta(x, y, t)$ together with boundary and initial conditions.

It would be possible to derive again a potential equation for $\varphi(x, y, z, t)$ like (3.5.29), but an exact linearization seems not to be feasible since methods to linearize nonlinear partial differential equations work mainly for two independent vriables only. Specializing for two independent variables we loose the vertical motion $v_z$. The problem of a multidimensional time dependent water flow over a locally varying bottom $h(x, y)$ is an extremely difficult nonlinear mathematical problem. Many attempts to solve such problems may be found in the literature. The problem to solve the basic nonlinear partial differential equations together with initial conditions is furthermore connected with geophysics since the initial conditions are realized by unpredictable earthquakes. Readers more interested in the tsunami problem will find specialized publications in three different areas:

1. attempts to find mathematical tools [3.14], [3.16],
2. work related with earthquakes [3.17],
3. solution methods, linearization, series expansions [3.13], graphics etc [3.18]

Research on tsunamis may also be found in the internet [3.19].

We now will derive the basic equations for a *one-dimensional time dependent tsunami* described by the crest height $\eta(x, t)$ and flow speed $u(x, t)$, which propagates into the $x$-direction. The reader should consider the figure 3.5. $\eta(x, t)$ designates the local tsunami elevation over the $x$-axis defined by $z = 0$. The local depth is again measured by $h(x)$. Then the *continuity equation* (3.5.16) reads

$$\eta_t + u\eta_x + u_x\eta + h_x u + u_x h = 0, \tag{3.5.61}$$

compare with (3.1.4). This corresponds to the neglection of velocity components in other directions than $x$. The hydrostatic pressure is now given by (3.1.2)

$$p(x, t) = \rho_0 g(\eta(x, t) - z). \tag{3.5.62}$$

Then the *equation of motion* (3.5.7) reads

$$u_t + u u_x + g\eta_x = 0, \tag{3.5.63}$$

compare with (3.3.1). The equations (3.5.61) and (3.5.63) represent a *shallow water* wave approximation for the tsunami [3.11]. The equations (3.5.61) and (3.5.63) are two quasilinear partial differential equations of first order. Reading off from (1.4.18) and substituting $v \to \eta$, $y \to x$, $x \to t$ we find the following coefficients from (1.4.18)

$$\begin{aligned} &a_{11} = 0, \quad a_{12} = 1, \quad b_{11} = \eta + h, \quad b_{12} = u, \quad h_1 = -h_x u, \\ &a_{21} = 1, \quad a_{22} = 0, \quad b_{21} = u, \qquad\quad b_{22} = g, \quad h_2 = 0. \end{aligned} \tag{3.5.64}$$

Then the condition of the vanishing determinant $R$ (1.4.23) yields

$$k' = \frac{\mathrm{d}x}{\mathrm{d}t} = u \pm \sqrt{g(\eta + h)} = u \pm c, \tag{3.5.65}$$

compare with (2.10.31), (3.5.28). Equation (3.5.65) describes the *propagation speed of the tsunami.* With increasing flow velocity $u$ and the increasing crest height $\eta$ the tsunami propagation speed increases. Now we investigate the compatibility conditions (1.4.24) and (1.4.25). Using (3.5.64) and inserting from (3.5.65) we obtain from (1.4.24) and (1.4.25)

$$V_1 = \pm\sqrt{g(\eta + h)}\,(h_x u + \eta_t) - u_t(h + \eta) \tag{3.5.66}$$

and

$$V_2 = \pm\sqrt{g(\eta + h)}u_t - g h_x u - g\eta_t, \tag{3.5.67}$$

and integration of $V_2 = 0$ yields (3.5.99). These two equations describe the modification of the *state variables* $u, \eta$ along the characteristics (3.5.65). One now could think to use the compatibility condition (3.5.67) to again establish RIEMANN *invariants* $r, s$, see (2.10.34). But there is fundamental difference between the compatibility condition (2.10.32) and (3.5.67). The latter contains a term $(g h_x u)$, which is not a differential. If we could define RIEMANN invariants then the $u, \eta$ or better $r, s$ plane could be called the principal plane or *state plane*, whereas the $x, t$ plane is usually defined as the *physical plane.* We cannot follow the method described in section 2.10. The distinction is that the depth $h(x)$ is no longer constant, but a given variable and may describe e.g., an inclined ocean bottom line

$$h(x) = ax + b, \quad h(0) = b = 4000\,\mathrm{m}, \tag{3.5.68}$$

or other functions. Now the initial conditions are very important. We might assume that a local earthquake at $x_0 = 0$ initiates at time $t = 0$ a sharp, but not very high elevation $\eta(0,0)$ in the form of a slender *soliton* or a GAUSSIAN error function with a crest height of 0.5 m. For the initial profile we could use

$$\eta(x,t) = A \cdot \text{sech}^2\left(B[x - c_o t]\right) \tag{3.5.69}$$

and have the *initial condition*

$$\eta(x,0) = A\text{sech}^2(B_x) \quad \text{or} \quad \eta(0,0) = A, \tag{3.5.70}$$

which is shown in Fig. 3.6. We may assume $\eta_1 = \eta(0,0) = A = 0.5$ m.

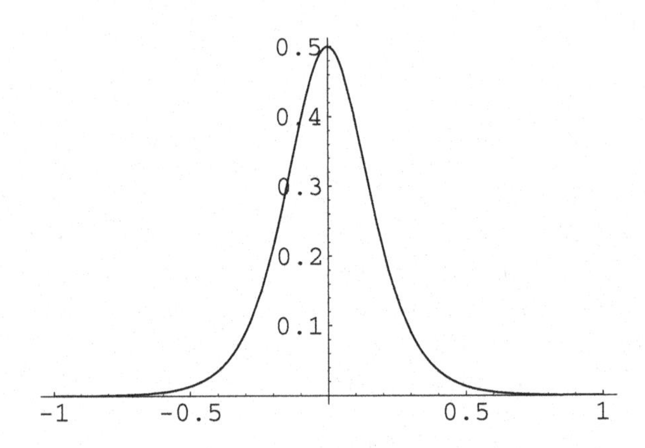


Fig. 3.6. Initial condition

This figure has been produced by the Mathematica commands:

```
Clear[A,B,y];A=0.5;B=5.;
y[x_]=A*(Sech[B*x])^2;                    (3.5.71)
Plot[y[x],{x,-1.,1.}]
```

In order to use again RIEMANN invariants we now transform (3.5.61) and (3.5.63). We use [3.11] the local propagation speed $c$ according to (3.5.65) in the form

$$c = \sqrt{g(\eta(x,t) + h(x))}, \quad \eta = \frac{c^2 - hg}{g}, \tag{3.5.72}$$

compare with (3.2.22). Calculating

$$c_x = \frac{g}{2c}\left(\eta_x + h_x\right), \quad c_t = \frac{g\eta_t}{2c} \tag{3.5.73}$$

we insert $\eta_x + h_x$ and $\eta_t$ into the equation of continuity (3.5.61) to obtain

$$2c_x u + 2c_t + u_x c = 0. \tag{3.5.74}$$

The equation of motion (3.5.63) becomes

$$u_t + uu_x + 2c_x c - gh_x = 0. \tag{3.5.75}$$

Now we again use the equations (1.4.18) and (1.4.23) - (1.4.25). In order to bring (3.5.74) and (3.5.75) into the form (1.4.18) we have to make the following substitutions $v \to c$, $x \to t$, $y \to x$ which give

$$\begin{aligned} &a_{11} = 1, \quad a_{12} = 0, \quad b_{11} = u, \qquad b_{12} = 2c, \quad h_1 = gh_x, \\ &a_{21} = 0, \quad a_{22} = 1, \quad b_{21} = c/2, \quad b_{22} = u, \qquad h_2 = 0, \end{aligned} \tag{3.5.76}$$

which does not coincide with (3.5.64). Then $R = 0$ according to (1.4.23) yields

$$k' = \frac{\mathrm{d}x}{\mathrm{d}t} = u \pm c, \tag{3.5.77}$$

where $c$ is defined by (3.5.72), compare (3.5.65). The two comparability conditions (1.4.23) and (1.4.24) are then given by

$$2V_1 = c(+2c_t + u_t - gh_x), \tag{3.5.78}$$

$$V_2 = c(-2c_t + u_t - gh_x). \tag{3.5.79}$$

Since $gh_x = ag$ and $(\partial/\partial t)(agt) = ag$, one concludes that [3.11]

$$u + 2c - agt = r = \text{const} \quad \text{for} \quad \frac{\mathrm{d}x}{\mathrm{d}t} = u + c, \text{(downstream)}, \tag{3.5.80}$$

or

$$u - 2c - agt = s = \text{const} \quad \text{for} \quad \frac{\mathrm{d}x}{\mathrm{d}t} = u - c, \text{(upstream)} \tag{3.5.81}$$

are constant along the two characteristics. For a downstream wave $r$ is constant, but $s$ varies. In analogy to (2.10.34) we introduce the RIEMANN *invariants* $r$ and $s$ so that

$$u = \frac{r+s}{2} + agt, \quad c = \frac{r-s}{4}. \tag{3.5.82}$$

Now we assume the initial and final conditions

$$\begin{aligned} &t = 0, \quad x = 0, \qquad\quad \eta_0 = 0.5\,\text{m}, \quad h_0 = 4000\,\text{m}, \quad c \approx 198\,\text{m}\,\text{s}^{-1}, \quad u_0 = 1, \\ &t = ?, \quad x = 300\,\text{km}, \quad \eta = 30\,\text{m}, \quad h = 5\,\text{m}, \qquad\quad c \approx 18\,\text{m}\,\text{s}^{-1}, \quad u = ?. \end{aligned} \tag{3.5.83}$$

Assuming a distance $\Delta x$ of 300 km between $x = 0$ (location of earthquake) and the coast, (3.5.68) gives $a = -0.0133167$ and $ag = -0.130637\ \mathrm{m\,s^{-2}}$. Thus the initial values $r_0$ and $s_0$ are

$$r_0 = 2\sqrt{g(4000 + 0.5)} + 1\,\mathrm{m\,s^{-1}}, \quad s_0 = -395.206537. \tag{3.5.84}$$

The final state values $r_N$ and $s_N$ cannot be given, since $r$ and $s$ now depend on $t$ as a consequence of the varying ocean depth $h(x)$. But a guess may be allowed in order to find a final $s$. If the tsunami arrives the coast at an ocean final depth $h_f = 5$ m and the final crest height $\eta_f$ is 30 m, then (3.5.65) yields $c_f = \sqrt{g(30 + 5)} = 18.5$, see (3.5.83). On the other hand (3.5.82) and $r_f = r_0 = 397.20$ allow to calculate $s_f$. This gives $s_f = +323.20$. If one is satisfied by $N = 100$ steps, then

$$\Delta s = s_0 - s_f = 718.40, \quad \delta s = \delta s/N = 7.184. \tag{3.5.85}$$

Finally, we need a step in time or location. Since we do not know the arrival time $t_f$ of the tsunami at $x = 300$ km, we choose a step in $x$.

$$\Delta x = 300\,000, \quad \delta x = 3\,000\,\mathrm{m}. \tag{3.5.86}$$

Furthermore one has to observe stability criteria for the numerical procedure. The Courant-Friedrichs-Lewy *condition*

$$\Delta t = 0.9 \frac{\Delta x}{2(u + c)_{\max}} \tag{3.5.87}$$

ensures stability of calculation [3.23], [2.20].

The various modern numerical methods are however outside of the scope of this book. They should consider an arbitrary $h(x) \neq ax + b$.

The continuity equation (3.5.16) and the equation of motion (3.5.7) are two quasilinear partial differential equations of first order. We shall now derive a second-order partial differential equation from these two equations. The result may be termed as wave equation for a dissipationless tsunami.

To derive such a *tsunami wave equation* we again use a mass variable transformation. We define

$$\begin{aligned} m(x,t) &= \textstyle\int j(x,t)\mathrm{d}\tau, \qquad j(x,t) = \eta(x,t) + h(x), \\ m_x &= j = \eta + h, \qquad m_t = -uj. \end{aligned} \tag{3.5.88}$$

Then the continuity equation (3.5.16) takes the form

$$j_t + j_x u + u_x j = 0, \tag{3.5.89}$$

since $h_t = 0$. Using the well known relations (see the analogous equations (3.5.37) etc)

$$j_t = -\hat{j}_m \hat{u}\hat{j} + \hat{j}_t, \qquad j_x = \hat{j}_m \hat{j}, \qquad u_t = -\hat{}_m \hat{u}\hat{j} + \hat{u}_t, \qquad u_x = \hat{u}_m \hat{j} \tag{3.5.90}$$

we again obtain a new form for the continuity equation

$$\hat{j}_t + \hat{j}^2 \hat{u}_m = 0. \tag{3.5.91}$$

Using again a transformation

$$\hat{j} = \frac{1}{\hat{s}}, \quad \hat{j}_t = -\frac{1}{\hat{s}^2}\hat{s}_t, \quad \hat{j}_m = -\frac{1}{\hat{s}^2}\hat{s}_m \tag{3.5.92}$$

one obtains

$$\hat{u}_m - \hat{s}_t = 0, \quad \frac{\partial}{\partial m}(\hat{u}) = \frac{\partial}{\partial t}(\hat{s}), \tag{3.5.93}$$

compare with (3.5.46). Defining again a function $W(m,t)$ by

$$W_t(m,t) = \hat{u}(m,t), \quad W_m(m,t) = \hat{s}(m,t) = 1/\hat{j} \tag{3.5.94}$$

the continuity equation is satisfied.

Now we consider the equation of motion which may now be written in the form

$$u_t + u u_x + g j_x = 0. \tag{3.5.95}$$

After the $m$-transformation it reads

$$\hat{u}_t + g\hat{j}_m \hat{j} = 0. \tag{3.5.96}$$

The transformation (3.5.92) yields

$$\hat{s}^3 \hat{u}_t - g\hat{s}_m = 0. \tag{3.5.97}$$

Insertion of $\hat{s}$, $\hat{s}_m$ and $\hat{u}_t$ from (3.5.94) yields the *tsunami wave equation*

$$W_m^3 W_{tt} = g W_{mm}, \tag{3.5.98}$$

compare with (3.5.52) and (3.5.54).

If a numerical or analytical solution of this nonlinear wave equation were available, all tsunami variables $u(x,t), \eta(x,t)$ etc could be calculated.

## Problems

1. Follow calculations from (3.5.61) to (3.5.67) and form differentials using (3.5.68).

   Solution:
$$\pm \mathrm{d}u + \frac{\sqrt{g}}{\sqrt{\eta + h}}\mathrm{d}\eta - \frac{a\sqrt{g}}{\sqrt{\eta + h}}\mathrm{d}(ut) = 0. \tag{3.5.99}$$
   What conclusions have we to draw from this result? (Formerly we could make a transformation (2.10.33) $\eta \to q \to \lambda \to r, s$, but instead of the two leading variables $r, s$ we now have a foreign variable $t$ from another plane!) This means that the method of section 2.10 can not be used. (3.5.99) should be solved by numerical methods which are not within the scope of this book.

2. Use the continuity equation to prove $\mathrm{d}m/\mathrm{d}t = 0$. Hint: for a channel of width $B$ of constant depth the water surface will be given by $h(x, t)$. Then
$$m = B\rho_0 \int_x^{x+\Delta x} h(x, t)\mathrm{d}x.$$
   Use $\rho = B\rho_0 h$.

   Solution [3.15]
$$\frac{\mathrm{d}}{\mathrm{d}t}m(t) = B\rho_0 \frac{\mathrm{d}}{\mathrm{d}t}\int_x^{x+\Delta x} h(x, t)\mathrm{d}x = B\rho_0 \int_x^{x+\Delta x}\left(h_t + \frac{\partial}{\partial x}(uh)\right)\mathrm{d}x = 0.$$
   The integrand replaces the continuity equation.

3. The continuity equation (3.5.42) could be linearized using the transformation (3.5.45), see (3.5.46). What is the result of this transformation for the equation of motion (3.5.43)? Hint: use (3.5.49) and (3.5.41).

   Solution: the equation for $\hat{u}$ and $\hat{s}$ becomes heavily nonlinear.

4. Use a similarity transformation
$$U(x, t) = x^\alpha t^\beta \tag{3.5.100}$$
   to solve the nonlinear partial differential equation
$$\frac{\partial u}{\partial t} = \frac{\partial}{\partial x}\left(f(u(x, t))\frac{\partial u}{\partial x}\right) \tag{3.5.101}$$
   and to transform it into an ordinary differential equation for arbitrary $f(u)$. Hint:
$$u_x = \frac{\mathrm{d}u}{\mathrm{d}U}\frac{\partial U}{\partial x} = u'\alpha x^{\alpha-1}t^\beta \text{ etc}, \quad u' = \frac{\mathrm{d}u}{\mathrm{d}U}.$$

Solution:

$$\frac{\beta x^2}{t} U u' = \alpha(\alpha - 1) U f(u) u' + \alpha^2 U^2 \frac{\mathrm{d}}{\mathrm{d}U} (f(u) u'),$$

where you have the choice a) $x^2/t = U$, b) $x^2/t = U^2$ or $x/\sqrt{t} = U$.

5. A polytropic spherical wave in a gas is described by

$$\begin{aligned} &\rho_t + u\rho_r + \rho\left(u_r + \frac{2u}{r}\right) = 0, \\ &u_t + uu_r + \frac{1}{\rho} p_r = 0, \\ &\frac{\partial}{\partial t}\left(p\rho^{-n}\right) + u \frac{\partial}{\partial r}\left(p\rho^{-n}\right) = 0, \end{aligned} \tag{3.5.102}$$

where $\rho(r,t), u(r,t), p(r,t)$. Solve this system by *similarity transformation*

$$\begin{aligned} &u(r,t) = \alpha r t^{-1} U(\xi), \qquad p(r,t) = \alpha^2 r^{\kappa+2} t^{-2} P(\xi), \\ &\rho = r^{\kappa} \Omega(\xi), \qquad \xi = r^{-\lambda} t, \quad \lambda = 1/\alpha. \end{aligned} \tag{3.5.103}$$

Derive the nonlinear ordinary differential equation for $U(\xi), P(\xi)$ and $\Omega(\xi)$.

Solution for the first equation in (3.5.102):

$$\Omega' \xi - \xi(U\Omega' + U'\Omega) + U\Omega(3\alpha + \alpha\kappa) = 0.$$

6. An unsteady, one-dimensional polytropic gas is represented by the system of equations [2.13]

$$\rho u_t + \rho u u_x + p_x = 0, \tag{3.5.104}$$

$$\rho_t + (\rho u)_x = 0, \tag{3.5.105}$$

$$S_t + u S_x = 0, \tag{3.5.106}$$

$$S = S_0 + c_V \ln(V \rho^{-\kappa}), \tag{3.5.107}$$

see also (2.5.46). Multiply (3.5.105) by $u$ and add the result to (3.5.104). This gives

$$\rho_t u + u_t \rho + p_x + \rho_x u^2 + 2\rho u u_x = 0. \tag{3.5.108}$$

Now satisfy (3.5.105) by the setup

$$\rho = \varphi_{xx}, \quad \rho u = -\varphi_{xt} \tag{3.5.109}$$

and receive from (3.5.108)

$$(p + \rho u^2) = \varphi_{tt}. \tag{3.5.110}$$

Express $u$, $p$ and entropy $S$ by $\varphi$. Assume $S = f(\varphi_x)$ and derive the equation for $\varphi(x,t)$.

Solution:

$$\varphi_{xx}\varphi_{tt} - \varphi_{xt}^2 = \varphi_{xx}^{\kappa+1} \cdot F(\varphi_x), \tag{3.5.111}$$

where $F(\varphi_x) = \exp[(1/c_V)f(\varphi_x)]$.

7. Solve (3.5.111) by the *method of the unknown function* making the setup

$$\varphi(x,t) = x^m H(t) \tag{3.5.112}$$

and $F(\varphi_x) = K \cdot \varphi_x^n$, where $K$, $m$ and $n(n \neq 1)$ are constants. Integrate the second-order nonlinear ordinary differential equation for $H(t)$.

Solution [2.13] after a first integration

$$H'^2 \cdot H^{-2m/(m-1)} =$$

$$\left(2Km^{(m+\kappa)/(m-1)} \cdot (m-1)^{(\kappa+1)/(\kappa-1)}\right) \cdot \left(H^{(\kappa-1)/(m+-1)} + \beta\right)$$

and after a second integration

$$t\sqrt{2Km^{(m+\kappa)/(m-1)} \cdot (m-1)^{(\kappa+1)/(\kappa-1)}} =$$

$$(1-m)\int \left(g^{1-\kappa} + \beta\right)^{1/2} \mathrm{d}g, \quad g = H^{-1/(m-1)}.$$

$\beta$ is an integration constant. Discuss the solution for $\beta = 0$ and $\beta \lessgtr 0$.

8. Although a soliton may not be adequate to calculate the *tsunami energy density* $E$, (3.4.18) may allow to give an estimate. Calculate $\rho v^2$.

Solution:

$$E = \text{const}\,\rho_0 c_0^2 \cdot \text{sech}^4\left(\frac{\sqrt{c_0}}{2}(x - c_0 t)\right) \quad [\text{kg}\,\text{m}^{-1}\,\text{s}^{-2} \text{ or } \text{Jm}^{-3}]. \tag{3.5.113}$$

9. The speed $a$ in (3.5.51) and (3.5.52) has not yet been defined. Assume (3.5.19), $\hat{\rho} = \rho_0 \eta$ and

$$a^2 = \frac{\mathrm{d}\hat{p}}{\mathrm{d}\hat{\rho}}. \tag{3.5.114}$$

Calculate $a$.

Solution:

$$a^2 = g(\eta + h), \tag{3.5.115}$$

compare with (3.2.22).

10. Using an expression for $a^2$, one may rewrite the equations for $W(m,t)$. Thus (3.5.52) reads

$$W_{tt} = W_{mm} \cdot g(\eta + h) \tag{3.5.116}$$

and (3.5.54) becomes for $\kappa = 2$

$$-W_{tt} = W_{mm}^{-2} \cdot \frac{g}{2}(\eta + h). \tag{3.5.117}$$

Eliminate $\eta$ by expressing it by a derivation of $W$.

Solution: For (3.5.52), (3.5.116) one gets

$$W_{tt} + W_{mm}\left(\frac{1}{W_m \rho_0} + gh\right). \tag{3.5.118}$$

Here we used (3.5.47). The given boundary condition $h(x)$ or $h(m)$ enters into the equation. If one would use

$$\hat{\rho} = \rho_0(\eta + h) \tag{3.5.119}$$

see (3.5.18), then (3.5.116) becomes

$$W_{tt} = W_{mm} \cdot g\left(\frac{1}{\rho_0 W_m}\right) \tag{3.5.120}$$

and $h$ does not enter the wave equation. Now consider (3.5.117).

Solutions:

$$-W_{tt} = W_{mm}^{-2} \cdot \frac{1}{2}\left(\frac{1}{W_m \rho_o} + gh\right) \tag{3.5.121}$$

and

$$-W_{tt} = W_{mm}^{-2} \cdot \frac{g}{2} \cdot \frac{1}{\rho_0 W_m}. \tag{3.5.122}$$

11. Use (3.5.43), (3.5.46) and (3.5.2) to derive two partial differential equations for $\hat{u}$ and $\hat{s}$. Derive a second-order equation for $\hat{s}$ alone.

Solution:

$$\hat{u}_m - \hat{s}_t = 0, \quad \hat{u}_t + \rho_0 g/\hat{s} = 0, \quad \hat{s}_{tt} \cdot \hat{s}^2 = g\rho_0 \hat{s}_m. \tag{3.5.123}$$

12. Solve the nonlinear second-order partial differential equation for $\hat{s}(m,t)$ by a similarity transformation.

    Solution: $\hat{s} = U(\zeta)$, $\zeta = m^\alpha t^\beta$, gives $\alpha = 1$, $\beta = 1$ and

$$\zeta U'' U^2 = g\rho_0 U'. \tag{3.5.124}$$

13. Differential equations of the type (3.5.52) may be solved by *travelling wave* solutions. Prove that $W = U(\zeta)$, $\zeta = m - at$ is actually a solution of (3.5.52).

14. Transform the equations (3.5.46) and (3.5.43) by using

$$\hat{p}(m,t) = g\rho_0\hat{\eta}(m.t), \quad \hat{s} = 1/\hat{\eta}\rho_0 \tag{3.5.125}$$

    and choosing $\hat{u}(m,t) = \hat{K}_{mt}(m,t)$, $\hat{s}(m,t) = \hat{K}_{mm}(m,t)$ into a partial differential equation for $K(m,t)$.

    Solution: (3.5.46) is automatically satisfied and (3.5.43) together with (3.5.125) yields

$$\hat{K}_{mtt} \cdot \hat{K}^2_{mm} - g\hat{K}_{mmm} = 0. \tag{3.5.126}$$

15. Apply the mass-transformation on the three-dimensional continuity equation (3.5.20) in the form

$$\eta_t + \vec{v}\nabla\eta + \eta\nabla\vec{v} = 0, \quad \nabla\vec{v} = \operatorname{div}\vec{v}. \tag{3.5.127}$$

    Hint: use

$$\begin{aligned} & m(x,y,z,t) = \textstyle\int\eta(x,y,z,t)\mathrm{d}\tau, \quad \nabla m = \vec{\eta}, \\ & \frac{\partial m}{\partial t} = -\frac{\partial m}{\partial x}\frac{\mathrm{d}x}{\mathrm{d}t} - \frac{\partial m}{\partial y}\frac{\mathrm{d}y}{\mathrm{d}t} - \frac{\partial m}{\partial x}\frac{\mathrm{d}z}{\mathrm{d}t} = -\nabla m\vec{v}. \end{aligned} \tag{3.5.128}$$

    Solution:

$$\vec{\eta}_t + \vec{\eta}^2(\nabla\vec{v}) = 0. \tag{3.5.129}$$

    What would one get from (3.5.34) in the form

$$\rho\frac{\partial\vec{v}}{\partial t} + \rho\nabla\frac{v^2}{2} + \nabla p = 0, \quad p = \rho_0 g(\eta - z). \tag{3.5.130}$$

    Hint: $\nabla p = \nabla\hat{p} \cdot \hat{\eta}$.

16. In this section we derived various *tsunami wave equations* (3.5.50), (3.5.54), (3.5.98), (3.5.120), (3.5.122), (3.5.126). Test the possibility to solve these equations by using similarity variables
$$W(m,t) = W(\zeta),\ \zeta = m^\alpha t^\beta.$$

## *3.6 Wave equation for dissipative tsunamis*

It is well known that taking into account dissipative terms like viscosity or heat conduction avoids the otherwise infinite steepness and multiple-valued functions of hydrodynamics [2.16], [2.18], [3.5], [3.20]. Thus it seems to be of interest to derive a wave equation for dissipative tsunamis. Due to dissipative effects, entropy increases and no potential $\varphi(x,y,z,t)$ describing the flow velocity exists. Furthermore, a BERNOULLI *equation* can no longer be derived and the polytropic state equation is no longer valid. The calculations have to be based on an equation of continuity, an equation of motion and on the energy theorem. Instead of a pressure-density relation $p(\rho)$ an expression for the hydrostatic pressure has to be used and it is doubtful if the equivalence theorem will lead to useful results.

We now investigate the simple case of a one-dimensional dissipative wave in an incompressible fluid of density $\rho_0$. For a water prism with a base in the bottom surface $x, y$ and varying height $\eta(x,t)$, see Fig. 3.5, the *continuity equation* (3.5.16) reads

$$\begin{aligned} &\frac{\partial}{\partial t}(\eta) + \frac{\partial}{\partial x}((\eta + h)u) = 0, \qquad && \eta + h = j, \\ &\frac{\partial}{\partial t}(j) + \frac{\partial}{\partial x}(ju) = 0, && h_t = 0. \end{aligned} \tag{3.6.1}$$

A derivation has been given in (3.1.4). This equation mainly takes into account the time variation of the height of the water column over the varying bottom $z = -h(x)$. The one-dimensional *equation of motion* including viscosity reads now

$$u_t + uu_x + gj_x - \nu u_{xx} = 0, \tag{3.6.2}$$

see (2.5.36). In the derivation we used $p(x,t) = \rho_0 g j(x,t)$, $\rho = \rho_0 j$ and $\nu$ is the kinematic viscosity, see section 2.1. The *energy theorem* (2.5.38) takes now the form

$$j\left(c_V T_t + c_V u T_x + u u_t + u^2 u_x\right) + g(u_x j + j_x u) - \nu u u_{xx} - \nu u_x^2 - \frac{\lambda}{\rho_0} T_{xx} = 0. \tag{3.6.3}$$

Here we used $U = c_V T$, see (2.5.40). We now have three nonlinear partial differential equations for $j(x,t), u(x,t)$ and $T(x,t)$.

As a **first step** we again make a transformation using the mass variable $m(x,t)$. For the convenience of the reader we rewrite the formulae (3.5.36) etc.

$$m(x,t) = \int_{x_0}^{x} j(x,t)\mathrm{d}x, \tag{3.6.4}$$

$$m_x = j(x,t), \quad m_t = -ju. \tag{3.6.5}$$

Functions depending on $m$ and $t$ take again a caret (ˆ). We have for the derivatives

$$\begin{aligned} j_t &= \hat{j}_m \cdot m_t + \hat{j}_t = -\hat{j}_m \hat{j}\hat{u} + \hat{j}_t, \\ u_t &= \hat{u}_m \cdot m_t + \hat{u}_t = -\hat{u}_m \hat{j}\hat{u} + \hat{u}_t, \\ T_t &= \hat{T}_m \cdot m_t + \hat{T}_t = -\hat{T}_m \hat{j}\hat{u} + \hat{T}_t, \end{aligned} \tag{3.6.6}$$

and also

$$\begin{aligned} j_x &= \hat{j}_m \cdot m_x = \hat{j}_m \cdot \hat{j}, \\ u_x &= \hat{u}_m \cdot m_x = \hat{u}_m \cdot \hat{j}, \\ u_{xx} &= \hat{u}_{mm} \cdot m_x^2 + \hat{u}_m \cdot \hat{j}_m = \hat{u}_{mm} \cdot \hat{j}^2 + \hat{u}_m \hat{j}_m \cdot \hat{j}, \\ T_x &= \hat{T}_m \cdot m_x = \hat{T}_m \cdot \hat{j}, \\ T_{xx} &= \hat{T}_{mm} \cdot m_x^2 + \hat{T}_m \cdot \hat{j}_x = \hat{T}_{mm} \hat{j}^2 + \hat{T}_m \hat{j}_m \hat{j}. \end{aligned} \tag{3.6.7}$$

Insertion into (3.6.1) gives the transformed continuity equation

$$\hat{j}_t + \hat{j}^2 \hat{u}_m = 0 \tag{3.6.8}$$

and the equation of motion (3.6.2) becomes

$$\frac{\partial}{\partial t}\hat{u} + \hat{j}\frac{\partial}{\partial m}\left(g\hat{j} - \nu \hat{u}_m \hat{j}\right) = 0. \tag{3.6.9}$$

Finally, the energy theorem (3.6.3) yields

$$\frac{\partial}{\partial t}\left(c_V \hat{T} + \frac{\hat{u}^2}{2}\right) + \frac{\partial}{\partial m}\left(g\hat{u}\hat{j} - \nu \hat{u}_m \hat{u}\hat{j} - \frac{\lambda}{\rho_0}\hat{T}_m \hat{j}\right) = 0. \tag{3.6.10}$$

For water one may assume $\rho_0 = 1$.

The **second step** consists of the transformation (2.2.7) in the form

$$
\begin{aligned}
&\jmath(x,t) = 1/s(x,t), && \hat{\jmath}(m,t) = 1/\hat{s}(m,t), \\
&\hat{\jmath}_t = -\hat{s}_t/\hat{s}^2, && \hat{\jmath}_m = -\hat{s}_m/\hat{s}^2.
\end{aligned}
\tag{3.6.11}
$$

This transforms the continuity equation (3.6.8) into a linear equation

$$\hat{u}_m - \hat{s}_t = 0 \tag{3.6.12}$$

and the equation of motion (3.6.9) yields

$$\frac{\partial}{\partial t}(\hat{u}) + \frac{1}{\hat{s}}\frac{\partial}{\partial m}\left(-\nu \hat{u}_m \frac{1}{\hat{s}} + \frac{g}{\hat{s}}\right) = 0. \tag{3.6.13}$$

The **third step** shall introduce a new function. We define $S(m,t)$ by

$$S_m = \hat{s}(m,t) = 1/\hat{\jmath}(m,t), \quad S_t = \hat{u}(m,t), \tag{3.6.14}$$

so that $\hat{u}_m = S_{tm}$ and $\hat{s}_t = S_{mt}$ satisfy the continuity equation (3.6.12).
Furthermore we have

$$\hat{u}_t = S_{tt}, \quad \hat{u}_m = S_{tm}, \quad \hat{u}_{mm} = S_{tmm}, \quad \hat{s}_m = S_{mm}. \tag{3.6.15}$$

Inserting into the equation of motion (3.6.13) one obtains a nonlinear partial differential equation of third order for $S(m,t)$

$$S_m^3 S_{tt} + S_{mm}\left(\nu S_{tm} - g\right) - \nu S_m S_{tmm} = 0. \tag{3.6.16}$$

For $\nu \to 0$, this equation is identical with (3.5.98). (3.6.16) replaces the continuity equation which is of first order and the equation of motion which is of second order. It may be called a *tsunami wave function* for a viscous, but not heat conducting fluid.

The **fourth step** considers the full energy theorem (3.6.10). Introduction of a new function $V(m,t)$

$$V_m = c_V \hat{T} + \frac{\hat{u}^2}{2}, \quad V_t = g\hat{u}\hat{\jmath} - \nu \hat{u}_m \hat{u}\hat{\jmath} - \frac{\lambda}{\rho_0}\hat{T}_m \hat{\jmath} \tag{3.6.17}$$

immediately satisfies (3.6.10) due to $V_{mt} = V_{tm}$. Since (3.6.10) is of second order, a combination with (3.6.16) which is of third order should yield a system of partial differential equations of fifth order describing a tsunami in a viscous and heat-conducting fluid.

From (3.6.17) we obtain

$$\begin{aligned}\hat{T} &= \left(V_m - \frac{\hat{u}^2}{2}\right)\frac{1}{c_V} = \left(V_m - \frac{1}{2}S_t^2\right)\frac{1}{c_V},\\ \hat{T}_m &= \frac{1}{c_v}\left(V_{mm} - S_t S_{tm}\right).\end{aligned} \tag{3.6.18}$$

Insertion into (3.6.17) yields the partial differential equation for $V(m,t)$ in the form

$$S_m V_t = S_t\left(g + S_{tm}\left(\frac{\lambda}{\rho_0 c_V} - \nu\right)\right) - \frac{\lambda}{\rho_0 c_V}V_{mm}. \tag{3.6.19}$$

Having solved (3.6.16) for $S(m,t)$ one may solve (3.6.19) for $V(m,t)$. Since (3.6.16) is of third order and (3.6.19) is of second order we have a system of fifth order. If a solution of (3.6.16) has been found, it is necessary to find the inverse functions. Since

$$\frac{\mathrm{d}\hat{x}}{\mathrm{d}t} = \hat{u}, \quad \hat{x}(m,t) = \int \hat{u}\mathrm{d}t + f(m), \tag{3.6.20}$$

where $f(m)$ is an arbitrary function to be determined by a boundary condition. Due to (3.6.14) one also has

$$\hat{x} = \int \hat{u}\mathrm{d}t = \int S_t\mathrm{d}t = S(m,t). \tag{3.6.21}$$

On the other hand, (3.6.5) and (3.6.11) demonstrate

$$\frac{\partial m}{\partial x} = \hat{\jmath}(m,t), \quad \frac{\partial x}{\partial m} = \hat{s}(m,t), \tag{3.6.22}$$

which yields

$$\hat{x}(m,t) = \int \hat{s}(m,t)\mathrm{d}m = \int S_m\mathrm{d}m = S(m,t) + g(t), \tag{3.6.23}$$

where we used (3.6.14) and where $g(t)$ is an arbitrary function to be determined by an initial condition. The flow speed $\hat{u}(m,t)$ and the specific volume $\hat{s}(m,t)$ as well as the crest height $\hat{\jmath}(m,t)$ may be calculated by integrations using (3.6.15). Thus one has

$$\hat{u}(m,t) = S_t(m,t), \quad \hat{s}(m,t) = S_m(m,t) = 1/\hat{\jmath}(m,t). \tag{3.6.24}$$

Partial differential equations of higher order have a large solution manifold. There are many solutions which do not describe tsunamis. To test solutions of (3.6.16) we wrote the following Mathematica code:

```
(* Solve 3.6.16 *)Clear[EQ,S,a,b,c,d]
EQ=(D[s[m,t],{m,1}])^3*D[S[m,t],{t,2}]+
D[S[m,t],{m,2}]*ν*D[S[m,t],{t,1},{m,1}]-
g*D[S[m,t],{m,2}]-ν*D[S[m,t],{t,1},{m,2}]*D[S[m,t],{m,1}];

(* 1.solution *)Clear[S];b=1.;
S[m_,t_]=a*t^b+m;
Simplify[EQ];

(* 2.solution *)Clear[S,a,b,c];
S[m_,t_]=a*t*(Log[t]-1)+t^b;
Simplify[EQ];

(* 3.solution *)Clear[S,a,b,c];
S[m_,t_]=a*m+t;
Simplify[EQ];                                          (3.6.25)
```

These three solutions $at^b + m$, $at(\log t - 1) + t^b$ and $am + t$ are primitive and do not describe a tsunami. One also could play around with a program using similarity transformations. Such a program reads:

```
(* Similarity solution of (3.6.16) =
EQ=(D[s[m,t],{m,1}])^3*D[S[m,t],{t,2}]+
D[S[m,t],{m,2}]*nu*D[S[m,t],{t,1},{m,1}]-
g*D[S[m,t],{m,2}]-nu*D[S[m,t],{t,1},{m,2}]*D[S[m,t],{m,1}] *)
Clear[a,b,S,Z,U,Zt,Zm,St,Stt,Stt1,Stt2,Sm,SmZ,Smm,Smm1,Smm2,
Stm1,Stm2,Stmm1,Stmm2,Stmm,ERG,AS,A1S,A2S,A3S,
ZERG,StZ,SttZ,SmmZ,StmZ,StmmZ,g,nu];
Off[General::spell1];Off[Set::write];Off[General::spell];
S[m_,t_]=U[Z];Z[m_,t_]=m^a*t^b;
Zt[m,t]=D[Z[m,t],t];
Zm[m,t]=D[Z[m,t],m];
Sm[m,t]=D[S[m,t],Z]*Zm[m,t] /. m^a*t^b->Z[m,t];
SmZ[m,z]=a*m^(-1)*Z*U'[Z];
St[m,t]=D[S[m,t],Z]*Zt[m,t];
StZ=b*Z*t^(-1)*U'[Z];
Stt1[m,t]=D[(St[m,t]*Zt[m,t]),Z];
Stt2[m,t]=D[St[m,t],t];
```

```
Stt[m,t]=Stt1[m,t]+Stt2[m,t];
SttZ=(-1+b)*b*t^(-2)*Z*U'[Z]+b^2*t^(-2)*Z^2*U''[Z];
Smm1[m,t]=D[Sm[m,t]*Zm[m,t],Z];
Smm2[m,t]=D[Sm[m,t],m];
Smm[m,t]=Smm1[m,t]+Smm2[m,t];
SmmZ=(-1+a)*a*m^(-2)*Z*U'[Z]+a^2*m^(-2)*Z^2*U''[Z];
Stm1[m,t]=D[St[m,t]*Zm[m,t],Z];
Stm2[m,t]=D[St[m,t],m];
Stm[m,t]=Stm1[m,t]+Stm2[m,t];
StmZ=a*b*m^(-1)*t^(-1)*Z*U'[Z]+a*b*m^(-1)*t^(-1)*Z^2*U''[Z];
Stmm1[m,t]=D[Stm[m,t]*Zm[m,t],Z];
Stmm2[m,t]=D[Stm[m,t],m];
Stmm[m,t]=Stmm1[m,t]+Stmm2[m,t];
Simplify[Stmm[m,t]];
StmmZ=a*b*m^(-2)*t^(-1)*Z*((-1+a)*U'[Z]+
Z*((-1+3*a)*U''[Z]+a*Z*U'''[Z]));
AS[m,t]=Simplify[-g*Smm[m,t]];
A1S[m,t]=Simplify[Stt[m,t]*Sm[m,t]^3];
Simplify[Stt[m,t]*Sm[m,t]^3];
A2S[m,t]=Simplify[-nu*Stmm[m,t]*Sm[m,t]];
A3S[m,t]=Simplify[nu*Stm[m,t]*Smm[m,t]];
ERG=Simplify[AS[m,t]+A1S[m,t]+A2S[m,t]+A3S[m,t]];          (3.6.26)
```

Here we used nu instead of the greek letter $\nu$, since the TeX transformation of Mathematica codes does not like greek letters.

Another program assuming a travelling wave solution might also be of theoretical interest, but does neither present tsunami solutions. Such a program could read

```
(* Travelling wave solution of (3.6.16) *)
Clear[S,Sm,Smm,Smmt,St,Stt,Stm,Equ];Off[General::spell1];
S[m_,t_]=F[m-a*t] /. m-a*t-> w;
Sm=D[F[m-a*t],{m,1} ] /. m-a*t-> w;
Smm=D[F[m-a*t],{m,2}]  /. m-a*t-> w;
Smmt=Simplify[D[F[m-a*t],{m,2},{t,1}] ] /. m-a*t-> w;
St=Simplify[D[F[m-a*t],{t,1}] ]  /. m-a*t-> w;
Stt=D[F[m-a*t],{t,2}] /. m-a*t-> w;
Stm=Simplify[D[F[m-a*t],{m,1},{t,1}] ] /. m-a*t-> w;
Equ=Sm^3*Stt+Smm*(nu*Stm-g)-nu*Sm*Smmt
```

$\mathrm{a}^2\ \mathrm{F}'\ [\mathrm{w}]^3\ \mathrm{F}''\ [\mathrm{w}] + \mathrm{F}''\ [\mathrm{w}]\ (\text{-g - a nu }\mathrm{F}''\ [\mathrm{w}]) + \mathrm{a\ nu\ F}'\ [\mathrm{w}]\ \mathrm{F}^{(3)}\ [\mathrm{w}]$

```
DSolve[Equ==0,F[w],w]
```

Solve::tdep :
The equations appear to involve the variables to be solved for in an essentially non-algebraic way.

$$\text{DSolve}[\text{a}^2\ \text{F}'\ [\text{w}]^3\ \text{F}''\ [\text{w}] + \text{F}''\ [\text{w}]\ (\text{-g - a nu F}''\ [\text{w}]) + \text{a nu F}'\ [\text{w}]\ \text{F}^{(3)}\ [\text{w}] == 0,\ \text{F}[\text{w}],\ \text{w}] \tag{3.6.27}$$

```
(* If the local speed of the wave is given, this equation can
be solved numerically. One may put the velocity a equal to a
given numerical value or equal to a function of the flow speed,
see (2.9.25) or similar *)
```

It seems that Mathematica is unable to solve such nonlinear ordinary differential equation. Numerical methods should again be considered.

It is not asthonishing that the latest publications of the year 2006 concern various numerical methods to handle similar problems. Dam-break wavefronts, especially propagating through a channel of varying depth exhibit some similarity with shallow water tsunamis. But brand new and quite old methods to solve numerically such problems are available [3.24].

## Problems

1. Variational calculus can help to solve differential equations because direct methods to solve the variational problem connected with a differential equation may be easier (Gröbner [3.16], Holz [3.24]). If one designates by $\eta(x,t)$ the location of the water surface above a bottom of varying profile $h(x)$, then for a width equal to $B = 1$ of the fluid, the potential energy $U$ is given by

$$U = \frac{1}{2} g\rho_0 \int_0^a \eta^2(x,t)\mathrm{d}x \tag{3.6.28}$$

and the kinetic energy $T$ is given by

$$T = \frac{\rho_0}{2}\int_0^a \mathrm{d}x \int_\eta^{h(x)} (u^2+v^2)\,\mathrm{d}z = \frac{\rho_0}{6}\int_0^a (h(x)-\eta(x,t)) \cdot \tag{3.6.29}$$

$$\cdot \left[3u^2 + h_2'^2 u^2 + h'u(\eta_x u + \eta_t) + (\eta_x u + \eta_t^2)\right] \mathrm{d}x.$$

$a$ is the length of the fluid mass in the $x$-direction. In the derivation of (3.6.29) it has been assumed that the horizontal component $u(x,t)$ of the flow speed does not depend on the vertical coordinate $z$. Furthermore for $v(x,z,t)$ it has been assumed that the vertical component may be expressed by

$$v = \frac{h(x+u\mathrm{d}t) - h(x)}{\mathrm{d}t} = h'(x)u(x,t), \quad \text{for} \quad z = h \tag{3.6.30}$$

and by

$$v = \eta_x(x,t)u(x,t) + \eta_t(x,t), \quad \text{for} \quad z = \eta. \tag{3.6.31}$$

Applying HAMILTON's principle

$$\Omega = \int_{t_1}^{t_2} (T - U)\mathrm{d}t \tag{3.6.32}$$

derive the appartaining differential equations for $\eta(x,t)$ for $h(x) = h_0 = \text{const}$ and for $h(x) = h_0 + \alpha x$, $\alpha < 0$.

Hint: the introduction of a new function

$$w(x,t) = \int_0^x \eta(x,t)\mathrm{d}x \tag{3.6.33}$$

may help but is not necessary.

Solutions (Gröbner):

$$\begin{aligned}\Omega = \frac{\rho_0}{6}\int_{t_1}^{t_2}\mathrm{d}t\int_0^a \Bigg[&\frac{(3+h'^2)w_t^2 + h'w_{xx}w_t^2 + w_{xx}^2 w_t^2}{h - w_x} + \\ &h'w_t w_{xt} + 2w_t w_{xx} w_{xt} + (h - w_x)\, w_{xt}^2 - 3g w_x^2\Bigg]\mathrm{d}x,\end{aligned} \tag{3.6.34}$$

where $w(x,t)$ is subject to the two boundary conditions

$$w(0,t) = w(a,t) = 0. \tag{3.6.35}$$

2. Now treat the case of a plane bottom $h(x) = h_0$. Assume a shallow water condition, that means that third powers of $w$ may be neglected.

Solution:

$$\Omega = \frac{\rho_0}{6}\int_{t_1}^{t_2} at \int_0^a \left[\frac{3}{h_0}w_t^2 + h_0 w_{xt}^2 - 3gw_x^2\right] dx \tag{3.6.36}$$

and

$$\frac{3}{h_0}w_{tt} - h_0 w_{xxtt} - 3gw_{xx} = 0 \tag{3.6.37}$$

is the appartaining differential equation. If the shallow water approximation ($v << u, w_{xt}^2 \approx 0$) is assumed, then the local wave propagation speed is given by $\sqrt{gh_0}$.

3. Now treat the case $h(x) = h_0 + \alpha x$, $\alpha < 0$ for shallow water conditions. Neglecting the vertical component, the solution is

$$(3 + \alpha^2)w_{tt} - 3gh(x)w_{xx} = 0. \tag{3.6.38}$$

4. For large depth one may assume that the horizontal component $u$ varies with depth $z$. Assume

$$u(x, z, t) = \frac{h(x) - z}{h(x)} u_0(x, t), \tag{3.6.39}$$

$$u_0(x, t) = \frac{2h(x)}{(f(x) - \eta(x, t))^2}\int_0^x \eta_t(x, t) dx. \tag{3.6.40}$$

For the vertical componnent $v$ one may assume

$$v(x, z, t) = \frac{h'(z - \eta) + \eta_x(h - z)}{h - \eta} u + \frac{(h - z)^2}{(h - \eta)^2}. \tag{3.6.41}$$

Neglecting all higher powers and assuming $h = h_0$ ($\alpha = 0$) derive the EULER equation of the variational principle.

Solution:

$$\frac{4}{3}h_0 - \frac{1}{5}h_0 w_{xxtt} - gw_{xx} = 0. \tag{3.6.42}$$

5. To model tsunami propagation over large distances, the Earth's curvature must be taken into account. TITOV and GONZALES (NOAA Technical Memorandum ERL PMEL-112) proposed the MOST Model. Discusss the nonlinear shallow-water wave equations in spherical coordinates with a CORIOLIS term.

Hint: $\lambda$ is the geographical longitude, $\Phi$ is latitude, $d(\lambda, \Phi, t)$ is the undisturbed water depth, $h(\lambda, \Phi, t)$ is the water surface perturbation and $u(\lambda, \Phi, t)$ and $v(\lambda, \Phi, t)$ are the depth-averaged flow velocity components. $R$ is the Earth radius.

Solution:

$$h_t + \frac{(uh)_\lambda + (vh\cos\Phi)_\Phi}{R\cos\Phi} = 0, \tag{3.6.43}$$

$$\begin{aligned} u_t + \frac{uu_\lambda}{R\cos\Phi} + \frac{vu_\Phi}{R} + \frac{gh_\lambda}{R\cos\Phi} - \frac{gd_\lambda}{R\cos\Phi} - fv = 0, \\ v_t + \frac{uv_\lambda}{R\cos\Phi} + \frac{vv_\Phi}{R} + \frac{gh_\Phi}{R} - \frac{gd_\Phi}{R} + fu = 0. \end{aligned} \tag{3.6.44}$$

$f = 2\omega\sin\Phi$ is the CORIOLIS parameter, see section 4.2.

6. As an exercise for similarity solutions and for Mathematica, solve (3.6.43).

   Solution:

```
(* MOST.nb [3.2] Spherical coordinates and
   similarity transformations *)
Clear[ζ,a,b,c,d,e,f,g,U,V,H,ha,u,v,h,
ut,ul,uf,vl,vf,vt,ht,hl,hf,hat,hal,haf,Co];
ζ[t_λ_]=t^a*λ^b;g=f;g=d;e=-1;
u[ζ[t,λ],t]=U[ζ[t, λ],Φ]*t^c*λ^d;
v[ζ[t,λ],Φ]=V[ζ[t,λ], Φ]*t^e*λ^f;
h[ζ[t,λ],Φ]=H[ζ[t,λ], Φ]*λ^g;
ha[ζ[t,λ],Φ]=ha[ζ[t, λ],Φ];
ut=D[u[ζ[t,λ],Φ]*t^c*λ^d,t];
ul=D[u[ζ[t,λ],Φ]*t^c*λ^d,λ ];
uf=D[u[ζ[t,λ],Φ]*t^c*λ^d,Φ ];
vf=D[v[ζ[t,λ],Φ]*t^c*λ^d,Φ ];
vt=D[v[ζ[t,λ],Φ]*t^c*λ^d,t];
vl=D[v[ζ[t,λ],Φ]*t^c*λ^d,λ ];
ht=D[h[ζ[t,λ],Φ]*t^c*λ^d,t];
hl=D[h[ζ[t,λ],Φ]*t^c*λ^d,λ ];
hf=D[h[ζ[t,λ],Φ]*t^c*λ^d,Φ ];
hal=D[ha[ζ[t,λ],Φ],λ];
haf=D[ha[ζ[t,λ],Φ],Φ];
```

```
Co=ht+(ul*h+hl*u+vf*h*Cos[Φ]+v*hf*Cos[Φ]-
v*h*Sin[Φ])/(R*Cos[Φ])
```

$$c\ t^{-1+c}\ \lambda^{2\ d}\ H[t^a\ \lambda^b, \Phi] + a\ t^{-1+a+c}\ \lambda^{b+2d}\ H^{(1,0)}\ [t^a\ \lambda^b,\ \Phi] + \frac{1}{R}$$
$$(\mathrm{Sec}[\Phi]\ (\text{-}h\ v\ \mathrm{Sin}[\Phi] + t^c\ v\ \lambda^{2\ d}\ \mathrm{Cos}[\Phi]\ H^{(0,1)}\ [t^a\ \lambda^b, \Phi] +$$
$$h\ t^{-1+c}\ \lambda^{d+f}\ \mathrm{Cos}[\Phi]\ V^{(0,1)}\ [t^a\ \lambda^b,\ \Phi] +$$
$$u\ (2\ d\ t^c\ \lambda^{-1+2d}\ H[t^a\ \lambda^b,\ \Phi]+$$
$$b\ t^{a+c}\ \lambda^{-1+b+2d}\ H^{(1+0)}\ [t^a\ \lambda^b,\ \Phi]) +$$
$$h\ (2\ d\ t^{2c}\ \lambda^{-1+2\ d}\ U[t^a\ \lambda^b,\ \Phi] + b\ t^{a+2c}\ \lambda^{-1+b+2d}\ U^{(1,0)}\ [t^a\ \lambda^b,\ \Phi])))$$

```
(* Change of similarity variable *)
Clear[ζ,a,b,c,d,e,f,g,U,V,H,u,v,h,ut,ul,uf,vl,
vf,vt,ht,hl,hf,Co];
ζ[Φ_,λ_]=Φ^a*λ^b;c=0;e=-1;
u[ζ[Φ,λ],t]=U[ζ[Φ,λ ],t]*t^c*λ^d;
v[ζ[Φ,λ],t]=V[ζ[Φ,λ],Φ]*t^e*λ^f;
h[ζ[Φ,λ],t]=H[ζ[Φ,λ],Φ]*λ^g;
ut=D[u[ζ[Φ,λ],t]*t^c*λ^d,t];
ul=D[u[ζ[Φ,λ],t]*t^c*λ^d,λ];
uf=D[u[ζ[Φ,λ],t]*t^c*λ^d,Φ];
vf=D[v[ζ[Φ,λ],t]*t^c*λ^d,Φ];
vt=D[v[ζ[Φ,λ],t]*t^c*λ^d,t];
vl=D[v[ζ[Φ,λ],t]*t^c*λ^d,λ];
ht=D[h[ζ[Φ,λ],t]*t^c*λ^d,t];
hl=D[h[ζ[Φ,λ],t]*t^c*λ^d,λ];
hf=D[h[ζ[Φ,λ],t]*t^c*λ^d,Φ];
Co=ht+(ul*h+hl*u+vf*h*Cos[Φ]+v*hf*Cos[Φ]-
v*h*Sin[Φ])/(R*Cos[Φ])
```

$$\frac{1}{R}\Big(\mathrm{Sec}[\Phi]\ (\ \text{-}h\ v\ \mathrm{Sin}[\Phi] + v\ \lambda^{d+g}\ \mathrm{Cos}[\Phi]\ \ (H^{(0,1)}\ [\lambda^b\ \Phi^a,\ \Phi]) +$$
$$a\ \lambda^b\ \Phi^{-1+a}\ H^{(0,1)}\ [\lambda^b\ \Phi^a,\ \Phi]) +$$
$$u\,((d+g)\,\lambda^{-1+d+g}\,H[\lambda^b\,\Phi^a,\ \Phi] + b\,\lambda^{-1+b+d+g}\,\Phi^a\,H^{(0,1)}\,[\lambda^b,\ \Phi^a,\ \Phi]) +$$
$$h\ (2\ d\ \lambda^{-1+2\ d}\ U[\lambda^b\ \Phi^a,\ t] + b\ \lambda^{-1+b+2d}\ \Phi^a\ U^{(1,0)}\ [\lambda^b\ \Phi^a,\ t]) +$$
$$\frac{1}{t}(h\lambda^{d+f}\ \mathrm{Cos}\ [\Phi]\ (V^{(0,1)}\ [\lambda^b\ \Phi^a,\ \Phi] + a\ \lambda^b\ \Phi^{-1+a}\ V^{(1,0)}\ [\lambda^b\ \Phi^a,\ \Phi]))\Big)$$

Remark: The first choice for $\zeta$ gives $g = f = d$, $e = -1$ and the contradictions $c-1 = c = 2c$ and $f = d-1$. The second choice yields $c = 0$, $e = -1$, $d+g = d+g-1 = d+f$ and $t^{-1}$ as well $\Phi^{-1}$ remains. We have to conclude that similarity solutions are not feasible.

## 3.7 The tsunami wave equations

In sections 3.5 and 3.6 we derived several wave equations. Some of them contained a const or a wave speed and were discussed in problem 16 of section 3.5 with disappointing results. We now concentrate on a simple wave equation in two dimensions, space $m$ and time $t$. We start with the continuity equation (3.5.46) which we rewrite as

$$\hat{u}_m - \hat{s}_t = 0, \tag{3.7.1}$$

where $\hat{u}(m,t)$, $\hat{s}(m,t)$. $u(x,t) = \hat{u}(m,t)$ is the flow speed in the $x$-direction and $s(x,t) = \hat{s}(m,t) = 1/\hat{\eta}$ is defined by (3.5.125). $\eta(x,t) = \hat{\eta}(m,t)$ is the local distance between the ocean bottom and the ocean surface defined by $h(x)$. This definition has the advantages, that the profile $h(x)$ of the ocean bottom does not enter into the wave equation, but has to be considered in a boundary condition. The mass variable is again defined by (3.5.36) to (3.5.39).

Now we consider the equation of motion for a dissipationless tsunami, which is given by (3.5.43). We rewrite it as

$$\hat{u}_t + \hat{p}_m = 0 \tag{3.7.2}$$

which is a partial differential equation of first order. $p(x,t) = \hat{p}(m,t)$ is the hydrostatic pressure given by (3.5.125). With $\rho_0 = 1$, one has

$$\hat{p}(m,t) = g\hat{\eta}(m,t) = g/\hat{s}(m,t). \tag{3.7.3}$$

Insertion into (3.7.2) yields

$$\hat{u}_t + \frac{\partial}{\partial m}(g\hat{\eta}) = \hat{u}_t + \frac{\partial}{\partial m}(g/\hat{s}) = \hat{u}_t - \frac{g}{\hat{s}^2}\hat{s}_m = 0, \tag{3.7.4}$$

which is now a non-linear differential equation.

To solve the two equations (3.7.1) and (3.7.2), we make a *similarity setup*

$$z = m^a t^b, \tag{3.7.5}$$

where we define the new wave function $K(m,t) \to K(z)$ by

$$\begin{aligned} \frac{\partial^2 K(m,t)}{\partial m \partial t} &= K_{mt}(m,t) = \hat{u}(m,t) \\ \frac{\partial^2 K(m,t)}{\partial m^2} &= K_{mm}(m,t) = \hat{s}(m,t) = 1/\hat{\eta}(m,t). \end{aligned} \tag{3.7.6}$$

Thus we have

$$\hat{u}_m = K_{mtm} = \hat{s}_t = K_{mmt} \tag{3.7.7}$$

which satisfies the continuity equation (3.7.1). The equation of motion (3.7.4) now assumes the form

$$K_{mtt} - gK_{mmm}K_{mm}^{-2} = 0 \tag{3.7.8}$$

which may be called a *tsunami wave equation*, compare with (3.5.126). In order to transform this nonlinear partial differential equation of third order into an ordinary differential equation for $K(z)$ we use (3.7.5). Thus we have

$$\left.\begin{aligned}
K_m &= \frac{\partial K}{\partial m} = \frac{\mathrm{d}K}{\mathrm{d}z}\cdot\frac{\partial z}{\partial m} = K'am^{a-1}t^b = m^{-1}aK'z,\\
K_{mm} &= \hat{s} = K''a^2m^{2a-2}t^{2b} + K'a(a-1)m^{a-2}t^b =\\
&\quad m^{-2}(K''a^2z^2 + K'a(a-1)z) = m^{-2}A,\\
K_{mt} &= \hat{u} = K''am^{a-1}t^b\cdot m^a bt^{b-1} + K'am^{a-1}m^a btb^{b-1} =\\
&\quad m^{-1}t^{-1}(K''abz^2 + K'abz).
\end{aligned}\right\} \tag{3.7.9}$$

To be able to insert into the two partial differential equations we need higher derivatives of $K(m,t) = K(z)$. One obtains

$$\left.\begin{aligned}
K_{mmm} &= \hat{s}_m = m^{-3}\left(K'''a^3z^3 + K''\left(a^2(2a-2)z^2+\right.\right.\\
&\quad \left.\left.a^2(a-1)z^2\right) + K'a(a-1)(a-2)z\right) = m^{-3}B,\\
K_{mtt} &= \hat{u}_t = m^{-1}t^{-2}\left(K'''ab^2z^3 + K''\left(a(2b-1)z^2+\right.\right.\\
&\quad \left.\left.ab^2z^2\right) + K'ab(b-1)z\right) = m^{-1}t^{-2}C.
\end{aligned}\right\} \tag{3.7.10}$$

Inserting into (3.7.8) we obtain after multiplication by $m^3$

$$m^{-2}t^{-2}A^2C - gB = 0, \tag{3.7.11}$$

which indicates that the choice $a = -2, b = -2$ produces an ordinary differential equation $z^2CA^2 - gB = 0$ or

$$\begin{aligned}
&z^2\left(4K''z^2 + 6K'z\right)^2\left(-8K'''z^3 + 2K''z^2 - 12K'z\right) -\\
&g\left(-8K'''z^3 - 36K''z^2 - 24K'z\right) = 0, \quad g = 9.81\text{m s}^{-2}.
\end{aligned} \tag{3.7.12}$$

If initial conditions at $z = c$ are given for $K(c), K'(c)$ and $K''(c)$, equation (3.7.12) can be integrated numerically using the Mathematica command `NDSolve`.

On the other hand, the equation (3.7.12) can be simplified by the modified setup defining $U(m,t) = U(z)$

$$U_t = \hat{u}, \quad U_m = \hat{s} \tag{3.7.13}$$

which replaces (3.7.6). It also satisfies (3.7.1). Instead of (3.7.8) we now have

$$U_m^2 U_{tt} - gU_{mm} = 0. \tag{3.7.14}$$

This corresponds to a formal integration of (3.7.8) with respect to $m$. But whereas (3.7.8) could be transformed into an ordinary differential equation, (3.7.14) seems to resist to a similarity transformation of the type (3.7.5).

Now we discuss dissipative tsunamis. The continuity equation (3.7.1) does not change, but the equation of motion does. Neglecting heat conduction effects which do not enter into the equation of motion and which only appear in the energy theorem we may use

$$u_t + uu_x + p_x - \nu u_{xx} = 0, \tag{3.7.15}$$

which transforms into

$$\hat{u}_t - \frac{g}{\hat{s}^2}\hat{s}_m - \nu\hat{u}_{mm}\frac{1}{\hat{s}^2} + \nu\frac{\hat{u}_m\hat{s}_m}{\hat{s}^3} = 0. \tag{3.7.16}$$

We now again define a function $K(m,t) = K(z)$ by the *similarity setup* (3.7.9) etc.. Then (3.7.16) yields

$$K_{mtt}K_{mm}^3 - gK_{mm}K_{mmm} - \nu K_{mm}K_{mtmm} + \nu K_{mtm}K_{mmm} = 0. \tag{3.7.17}$$

For $\nu = 0$ one comes back to (3.7.8). It seems that (3.7.17) resists to a similarity transformation of the type (3.7.5), as other equations exhibiting a third power term do. The cause for this behavior seems to be the term $u_{xx}$.

We now consider a numerical solution of (3.7.12). If $K_{mt}$ and $K_{mm}$ are known, then $\hat{u}(m,t)$ and $\hat{\eta}(m,t)$ are also known. Then $u(x,t)$ and $\eta(x,t)$ can be calculated from the following considerations. According to (3.5.37) one has

$$\frac{\mathrm{d}m(x,t)}{\mathrm{d}t} = \frac{\partial m}{\partial x}\frac{\mathrm{d}x}{\mathrm{d}t} + \frac{\partial m}{\partial t} = 0, \quad \left(\frac{\partial m}{\partial x}\right)_t = \eta(x,t), \tag{3.7.18}$$

so that again

$$\frac{\partial m}{\partial t} = -\frac{\partial m}{\partial x}\frac{\mathrm{d}x}{\mathrm{d}t} = -\eta(x,t)u(x,t) \tag{3.7.19}$$

is valid. Now for any function $F(x,t) = \hat{F}(m(x,t),t)$, be it $u(x,t) = \hat{u}(m(x,t),t)$, $\eta(x,t)$ or $x = \hat{x}(m,t)$, the following formulae are valid [2.18]:

$$\frac{\partial F(x,t)}{\partial x} = \frac{\partial \hat{F}(m(x,t),t)}{\partial m}\frac{\partial m}{\partial x}, \qquad \frac{\partial F}{\partial t} = \frac{\partial \hat{F}}{\partial m}\frac{\partial m}{\partial t} = -\hat{F}_m \eta u + \frac{\partial \hat{F}}{\partial t}. \quad (3.7.20)$$

Furthermore insertion into the total differential

$$\mathrm{d}F/\mathrm{d}t \quad \text{for} \quad F(x,t) = \hat{F}(m(x,t),t)$$

yields

$$\frac{\mathrm{d}F}{\mathrm{d}t} = \left(\frac{\partial \hat{F}}{\partial t}\right)_m = \frac{\partial F}{\partial x}\eta + \frac{\partial F}{\partial t}. \quad (3.7.21)$$

If one inserts $x = \hat{x}(m,t)$ into $F$, one obtains

$$F(\hat{x}(m,t),t) = \hat{F}(m,t). \quad (3.7.22)$$

This results in

$$\hat{F}_m = F_x \hat{x}_m = F_x/\eta, \quad \hat{F}_t = F_x \hat{x}_t + F_t = F_x u + F_t. \quad (3.7.23)$$

Then from (3.7.20) one has

$$\frac{\partial m}{\partial x} = \eta = \frac{1}{\hat{x}_m}. \quad (3.7.24)$$

Now let us assume that a solution

$$\hat{u} = \varphi_1(m,t), \qquad \hat{\eta} = \varphi_2(m,t) \quad (3.7.25)$$

is known and let the inverse functions be designated by $f_1$ and $f_2$, then

$$m = f_1(\hat{u},\hat{\eta}), \qquad t = f_2(\hat{u},\hat{\eta}). \quad (3.7.26)$$

Due to $\hat{x}_t = \hat{u}$, one has

$$\begin{aligned} \hat{x} &= \int \varphi_1(m,t)\mathrm{d}t = \int \hat{u}\mathrm{d}t, \\ \hat{x} &= \int \varphi_2(m,t)\mathrm{d}t = \int \frac{\mathrm{d}m}{\hat{\eta}}. \end{aligned} \quad (3.7.27)$$

On the other hand, if $f_1$ and $f_2$ are known, then $m = m(\hat{u},\hat{\eta})$, $t = t(\hat{u},\hat{\eta})$. Using

$$\Delta = \begin{vmatrix} m_u & t_u \\ m_\eta & t_\eta \end{vmatrix} \quad (3.7.28)$$

one may write [2.18]

$$\hat{u}_m = t_\eta/\Delta, \quad \hat{u}_t = -m_\eta/\Delta, \quad \hat{\eta}_m = -t_u/\Delta, \quad \hat{\eta}_t = m_u/\Delta. \tag{3.7.29}$$

Considering that $x$ and $t$ are functions of $u$ and $\eta$, one notes

$$u_x = y_\eta/D, \quad u_t = -x_\eta/D, \quad \eta_x = -t_u/D, \quad \eta_t = x_u/D, \tag{3.7.30}$$

where

$$D = \begin{vmatrix} x_u & t_u \\ x_\eta & t_\eta \end{vmatrix}. \tag{3.7.31}$$

The main problem is now, how to find the solutions from the ordinary differential equation (3.7.8). After neglection of integration constants (signifying a translation on the $x$-axis) we obtain from (3.7.26) and (3.7.27)

$$\begin{aligned} \hat{x} &= \int \hat{u}\mathrm{d}t = \int K_{mt}\mathrm{d}t = K_m; \quad \hat{x}_t = \hat{u}, \\ \hat{x} &= \int \frac{\mathrm{d}m}{\hat{\eta}} = \int K_{mm}\mathrm{d}m = K_m; \quad \hat{x}_m = 1/\hat{\eta}. \end{aligned} \tag{3.7.32}$$

The next necessary steps would be to solve first (3.7.8) and then transit

$$K(z(m,t)) \rightarrow K(m,t) \rightarrow K_{mt}(m,t) \rightarrow \hat{u}(m,t) \rightarrow u(x,t).$$

We use the following Mathematica command to solve (3.7.8) numerically

```
(* K2 Solve (3.7.12)
Integration starts at z=c with initial conditions d,e,f
 and ends at z=h *)
Clear[EQ,EQT,TA,z,K,K1,K2,K3,K4,KS1];
Off[NDSolve::ndnum];g=9.81;
EQ=z^2*(4*K''[z]*z^2 + 6*K'[z]*z)^2*(-8*K'''[z]*z^3+
2*K''[z]*z^2 - 12*K'[z]*z) -
g*(-8*K'''[z]*z^3 - 36*K''[z]*z^2 - 24*K'[z]*z);
Clear[c,d,e,f,h,z];c=0.5;d=0.1;e=0.01;f=0.001;h=10.;
z[m_,t_]=m^(-2)*t^(-2);
EQT=NDSolve[{EQ==0, K[c]==d, K'[c]==e, K''[c]==f}, K, {z,c,h}];
K1[z_]=Evaluate[K[z] /.%];
K2[z_]=Evaluate[K'[z] /.%%];
K3[z_]=Evaluate[K''[z] /.%%%];
K4[z_]=Evaluate[K'''[z] /.%%%%];

Plot3D[z[m,t],{m,c,h},{t,c,h},ColorOutput->None,PlotPoints->30]
```

This command generates Fig. 3.7. It is interesting, that the solutions are quite insensible to modifications of the initial conditions (c,d,e,f).

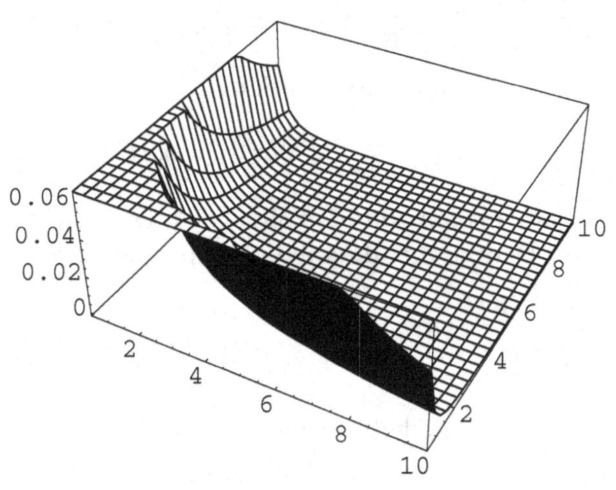


Fig. 3.7. Plot of $z(m,t)$

The command `Plot[Evaluate[K1{z],{z,c,h}]` delivers Fig. 3.8.

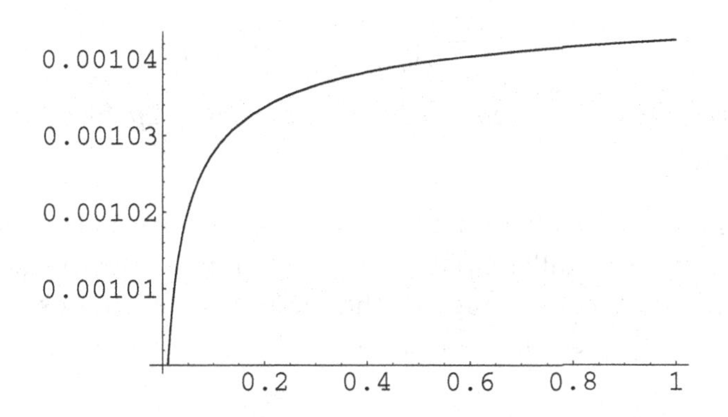


Fig. 3.8. Plot of $K1(z)$

Further work may concentrate on the definition of
`KS1{x_,t_]=Evaluate[K[x^(-2)*t^(-2)]/. n%`
on numerical integrations using the NIntegrate etc.

## Problems

1. The wave equation for $K(z)$ is based on the two equations (3.7.1) and (3.7.2). Prior to the mass-variable transformation they may be written

in the following form

$$u_t + uu_x + g\eta_x = 0, \tag{3.7.33}$$

$$\eta_t + u_x\eta + \eta_x u = 0. \tag{3.7.34}$$

This system has the similarity solution

$$u(x,t) = u((xt^{-1}), \qquad \eta(x,t) = \eta(xt^{-1}). \tag{3.7.35}$$

The Mathematica command `Plot3D[u[xt],{x, },{t }]` can produce a figure. Does $u(x,t)$ describe a steepening wave?

Solution: No. Why?

2. Solve (3.7.34) and

$$u_t + uu_x + g\eta_x - \nu u_{xx} = 0 \tag{3.7.36}$$

by the setup

$$\hat{x} + a^m x, \quad \hat{t} = a^r t, \quad \hat{u} = a^p u, \quad \hat{\eta} = a^s \eta.$$

Hint:

$$u_t = \frac{\partial u}{\partial \hat{u}} \frac{\partial \hat{u}}{\partial \hat{t}} \frac{\partial \hat{t}}{\partial t} = a^{r-p} \hat{u}_{\hat{t}}.$$

Solution: $-p + r = -2p + m = s + m = 2m - 2p$ for $p, r, m, s$ from (3.3.36).

3. The Lie *series method* [1.1] is also able to solve initial value problems of ordinary differential equations. We present two examples. To demonstrate the advantage of the method we choose examples with known closed analytical solutions.
Solve

$$F''(r) + \frac{1}{r}F'(r) - k^2 F(r) = 0. \tag{3.7.37}$$

Define

$$Z = \frac{\mathrm{d}F}{\mathrm{d}r}, \quad -\frac{1}{r} - Z + k^2 F = \frac{\mathrm{d}Z}{\mathrm{d}r} \tag{3.7.38}$$

and find

$$F(r) = Z_1(r), Z = Z_2(r) = \vartheta_1(Z_0, Z_1, Z_2), \; r = Z_0, \; \vartheta_2(r) = -\frac{1}{r}Z + k^2 F,$$

$$\frac{\mathrm{d}Z_0}{\mathrm{d}r} = \vartheta_0 = 1, \; \frac{\mathrm{d}Z_1}{\mathrm{d}r} = \vartheta_1(Z_0, Z_1, Z_2), \; \frac{\mathrm{d}Z_2}{\mathrm{d}r} = \vartheta_2(Z_0, Z_1, Z_2).$$

With the initial conditions $r = 0$, $Z_0(0) = 0$, $Z_1(0) = x_1$, $Z_2(0) = x_2$ the solution is

$$Z_i(r) = \sum_{\nu=0}^{\infty} \frac{r^\nu}{\nu!} \left(\mathrm{D}^\nu Z_i\right)_{r=0}, \tag{3.7.39}$$

where

$$\mathrm{D} = \sum_{\nu=0}^{\infty} \vartheta_\nu \left(Z_0, Z_1, \dots Z_n\right) \frac{\partial}{\partial Z_\nu}. \tag{3.7.40}$$

The reader might ask why this complicated sorcery since the BESSEL functions as solutions are very well known and tabulated numerically. There are, however, engineering problems exhibiting very large arguments ($\sim 50$) so that even large mainframes have problems with the numerical calculations. Now the LIE series method offers a way to avoid the BESSEL function by splitting the problem into

$$F_{\text{cyl}} = F_{\text{plane}} - F_{\text{correction}}. \tag{3.7.41}$$

The functions $\vartheta_\nu$ must be holomorphic (no singularities). Now $\vartheta_2$ has a pole at $r = 1$, so that a transformation is necessary. Let $d_\mu = r_\mu - r_{\mu-1}$ the thickness of the $\mu$-th radial domain, one may put $\xi = r - r_{\mu-1}$, where $\xi$ and $r$ are variables and $r_{\mu-1}$ is the radial distance of the $\mu$-th subdomain. Then

$$\frac{\mathrm{d}Z_0}{\mathrm{d}\xi} = 1 = \vartheta_0, \quad \frac{\mathrm{d}Z_1}{\mathrm{d}\xi} = Z_2 = \vartheta_1, \quad \frac{\mathrm{d}Z_2}{\mathrm{d}\xi} = \frac{Z_2}{r_{\mu-1} + \xi} + k^2 Z_1 = \vartheta_2 \tag{3.7.42}$$

and the new operator reads

$$\mathrm{D} = \frac{\partial}{\partial Z_0} + Z_2 \frac{\partial}{\partial Z_1} + \left\{k^2 Z_1 - \frac{Z_2}{r_{\mu-1} + Z_0}\right\} \frac{\partial}{\partial Z_2}. \tag{3.7.43}$$

The initial conditions are now $F_0 = F(\xi = 0), F'(0) = \mathrm{d}F(\xi = 0)/\mathrm{d}\xi$, so that the solution now reads

$$\begin{aligned} Z_1 \equiv F(\xi) &= F_0 \cosh k\xi + \frac{F_0'}{k} \sinh k\xi - \sum_{\nu=0}^{\infty} \xi^\nu \frac{f_\nu}{\nu!}, \\ &= F_{\text{plane}} \qquad\qquad\qquad - F_{\text{corr}}. \end{aligned} \tag{3.7.44}$$

Here $f_\nu(Z_0, Z_1, Z_2)$ is evaluated for $Z_0 = \xi = 0, Z_1 = F_0, Z_2 = F_0'$ and

$$f_\nu = k^2 f_{\nu-2} + \mathrm{D}^{\nu-2}\left[\frac{Z_2}{r_{\mu-1} + Z_0}\right]. \tag{3.7.45}$$

The operations can be calculated by recursion

$$\begin{aligned} \mathrm{D}^k\Big[\frac{Z_2}{r_{\mu-1}+Z_0}\Big] &= \frac{(l-2)k^2}{r_{\mu-1}+Z_0}\mathrm{D}^{l-3}\Big[\frac{Z_2}{r_{\mu-1}+Z_2}\Big]+ \\ k^2\mathrm{D}^{l-2}\Big[\frac{Z_2}{r_{\mu-1}+Z_0}\Big] &+ \frac{l+1}{r_{\mu-1}+Z_0}\mathrm{D}^{l-1}\Big[\frac{Z_2}{r_{\mu-1}+Z_0}\Big]. \end{aligned} \tag{3.7.46}$$

Since the analytic solution is known, one can compare.
For $F_0 = F_o' = 1$, $k = 0.1$, $r_{\mu-1} = 100, \mathrm{d} = 20$ one has

$$F_{\text{correction}} = F_{\text{plane}} - F_{\text{analytic}} =$$

$$= \cosh 2 + \frac{\sinh 2}{0.1} - 10\mathrm{K}_1(10)\mathrm{I}_0(12) + \mathrm{I}_1(10)\mathrm{K}_0(12) + \\ +100(\mathrm{K}_0(10)\mathrm{I}_0(12) - \mathrm{I}_0(10)\mathrm{K}_0(12) = 3.3\,667,$$

whereas the LIE solution up to $\mathrm{D}^4$ gives 3.3 664, I and K are the modified BESSEL functions.

4. Investigate if $t \log m$, $mt^2$, $t^{2/3}m^{-1}$, $m^2t^{2/3}$, $m^2+amt^2$ satisfy (3.7.8). For $K_m(-ct/m)$ one should obtain $u = -c/m$, $x = -ct/m, K = -ct\log m$, $K_{mt} = -c/m$.

# 4. Hurricanes

## 4.1 Terminology and basics

Various names are applied to rotating wind systems of a stormy character. Basically in meteorology one uses the term *cyclone* for an area of relatively low pressure which exhibits a revolving air circulation. In such a low-pressure area the wind circulates counterclockwise on the northern hemisphere and clockwise on the southern hemisphere. On the other hand, an area with a pressure higher than in the surroundings is called an *anticyclone.* The circulation of an anticyclone is clockwise in the northern hemisphere and counterclockwise in the southern hemisphere. This behavior will be explained in the next section.

Cyclones are connected with bad weather and rain. In meteorology cyclones are classified as *tropical cyclones*, also called *typhoons* or *hurricanes* and *extratropical cyclones.* These cyclones are generated by a strong vertical temperature gradient perpendicular to the wind flow speed. The CORIOLIS *force*, see section 4.2, delivers the necessary instability at latitudes greater than about 5 degrees. Tropical cyclones are caused by low air pressure areas and the CORIOLIS force. The US Atlantic Oceanographic and Meteorological Laboratory [4.1] defines some factors able to form a tropical hurricane:

1. Warm ocean water of at least 26.5° C. (Below this value the atmosphere is too stable).
2. An atmosphere which cools such that an unstable convection occurs. The thunderstorm itself allows the liberation of the heat stored in the ocean water. Heating from below may be up to 250 $\mathrm{W\,m^{-2}}$.
3. A minimum distance ($>$ 500 km) from the equator, so that the CORIOLIS force is strong enough.

Hurricanes can produce tornadoes and waterspouts.

In the proper original sense, a hurricane is a tropical storm occurring in the West Indies. These storms last roughly only in the months June to October, but have been observed not only in the Caribbean, the Gulf of Mexico or in the Pacific, but also on the northern coast of Australia (cyclone Larry with 290 $\mathrm{km\,h^{-1}}$) or near Hamburg.

Hurricanes need a pre-existing weather condition to be generated. Such conditions are mainly an ocean water surface temperature of about 26° C and pre-existing convective wave perturbations ("*easterly waves*"). Such waves are commonly embedded into trade winds. A hurricane is a tropical cyclone generated by a low pressure system and is accompanied by a thunderstorm. Severe cyclonic disturbances in low latitude are also called *tropical storm*. *Typhoon* is a local name of a tropical storm in the western North Pacific. Typhoons are distinguished from tropical storms in other parts of the world by their large number (~20 per year). They occur in all months of the calendar year. Tropical cyclones in the gulf of Bengale are called "Zyklon" and if occurring in Australia, the tropycal cyclone is sometimes named "Willy-willy".

If a tropical cyclone reaches winds of at least $17\,\mathrm{m\,s^{-1}}$ it is called a *tropical storm*. If winds reach $33\,\mathrm{m\,s^{-1}}$, then they are called hurricanes in the North Atlantic or Northeast Pacific, but typhoon in the Northwest Pacific Ocean, severe tropical cyclone in the Southwest Pacific, severe cyclonic storm in the North India Ocean and tropical cyclone in the Southwest Indian Ocean [4.1], [2.9], [4.2].

*Extratropical cyclones* are generated, if the horizontal wind increases with height due to a certain temperature gradient. In this case a *baroclinic instability* occurs. Such an instability is generated if two air layers – hot and cold – have a relative velocity, see section 4.4. In this situation, available potential energy is transformed into kinetic energy of the pressure gradient (PICHLER [4.4], CHARNEY 1947).

The factors exciting the baroclinic instability may be summarized:

1. meridional temperature gradients,
2. vertical shear of geostrophic wind,
3. variation of the CORIOLIS parameter with the geographic latitude.

*Tropical cyclones* are however not generated by this mechanism. A baroclinic condition occurs if the fluid density is not a sole function of pressure – no polytropic state exists and the *solenoidal term* in (2.6.9) does not vanish [4.9].

For an intense rotary storm of small diameter the term *tornado* is used. Such a tornado is always extended downward from the bottom of a convective heavy cloud like cumulonimbus. A broad funnel with small diameter near ground forms a narrow rope-like vortex. Humid air spirals inward and rises rapidly in the core, also called eye. An intensely whirling vortex extending from the cloud down to the water surface is called a waterspout ("Wasserhose", "Trombe"). The funnel consists of water vapor condensed because of lower pressure within the vortex. Passing inland waterspouts dissipate rapidly due to friction on the ground. Requisite conditions for tornado formation are thermodynamic instability and the occurrence of sufficient amounts of water vapor to produce thunderstorms. These thunderstorms are also connected with *squalls* ("Windstoß", "Bö"), a strong wind with sudden onset. Squall lines are lines along which a number of simultaneous squalls is generated. They form near cold fronts of extratropical cyclones.

Tropical cyclones [4.3] are different from tornadoes. While both are atmospheric vortices, they have little in common. Tornadoes have diameters of several hundred meters and are generated from a cumulonimbus cloud. A tropical cyclone has a far larger diameter of up to several hundred kilometers. Tornadoes require a vertical shear of horizontal winds and are mainly an over-land phenomenon, but tropical cyclones are a purely oceanic phenomenon.

Hurricanes have an *eye*, an inside circular area of relatively dry air which the strongest winds circulate around [4.4]. There is nearly no precipitation near the eye whose center represents the rotation axis of the hurricane, compare Fig. 2.3 in section 2.7. A detailed description of eye formation may be found in [4.4].

Two scales describe wind velocities. The BEAUFORT *scale* [4.5] connects a descriptive term like calm (BEAUFORT number 0), strong breeze at ~25 miles/hour (number 6), hurricane (number 12, more than 65 miles) and has 17 steps. Remark: 1 mile per hour = 1.609 $\mathrm{km\,h^{-1}}$ = 0.4470 $\mathrm{m\,s^{-1}}$. For hurricanes there exists the SAFFIR-SIMPSON *scale* which has 5 categories. Winds up to 39–73 miles per hour (mph) or 17–33 $\mathrm{m\,s^{-1}}$ are called a *tropical storm.* To earn the name hurricane, the wind must blow faster, see Table 4.

Table 4.1. SAFFIR-SIMPSON hurricane scale

| category | 1 | 2 | 3 | 4 | 5 |
|---|---|---|---|---|---|
| pressure in mb | < 24 | 21–28.50 | 28.4–27.9 | 27.8–27.17 | < 27.12 |
| wind speed $\mathrm{m\,s^{-1}}$ | 32–42 | 42–49 | 50–57 | 58–68 | > 61 |
| water surge m | 1.32 | 2.13 | 3.20 | 4.57 | > 5.49 |
| damages | small | building roofs | structural damage | major erosion | complete building |

In a storm surge the water surface is set into circular motion by counter clockwise winds. This motion pushes water inside the hurricane, inward toward the eye. It generates a convergence of water masses in the surface layer: "The greatest potential for loss of life related to a hurricane is from the storm surge" (B. Jarwinen, National Hurricane Center, [4.1]). By a water surge the mean water level can increase by 5 m and more. A shallow slope off the coast will thus allow a greater inundation land area.

Hurricanes may release much energy in two ways:

1. energy released by the condensation,

2. kinetic energy from the strong swirling.

On the other hand, there exist studies [4.6] establishing a quantitative upper bound on hurricane power as measured by maximum surface wind speed. Global climate models based on anthropogenetic warming show a substantial increase in the hurricane power and frequency of occurrence [4.7]. Such considerations are important since the actual money loss by storms is roughly proportional to the third power of the wind speed. The total power dissipated annually by tropical cyclones in the North Atlantic has doubled between 1930 and 2003. An average hurricane produces 1.5 cm/day of rain inside a circle of radius 665 km. This respresents about $2 \cdot 10^{16}$ cm$^3$/day rain or a condensation heat of $5.2 \cdot 10^{19}$ J/day or a power of $6 \cdot 10^{14}$ W which is about 200 times the world-wide electrical capacity [4.8]. After the 2004 Atlantic hurricane season the extent of devastation has been estimated to be $ 40 billion ($33 \cdot 10^9$ Euro). On the other hand the European Geophysical Union in their April 2006 meeting in Vienna, Austria, discussed the idea that the next ice-age would be due in the very near future [4.8]. It is a fact that the ocean responds to a hurricane by cooling of the sea surface temperature: Scientists of the University of Colorado predict an increase of the hurricane frequency for the period June – November 2006. For the insurance industry and for governments weatherprediction, warning strategies and evacuation plans are of considerable importance.

## *4.2 The excitation of vorticity in cyclones*

In section 2.6 we discussed the vorticity theorems by Crocco, Helmholtz and Thomson describing the conservation of vorticity. A similar circulation theorem like Thomson's is due to Bjerknes. It reads: "motions along isentropic surfaces do not change the circulation $\Gamma$" [4.9]. Another vorticity theorem is due to Ertel: the scalar product of vorticity and the gradient of the potential temperature divided by the fluid density is constant (Ertel *vorticity theorem* [4.14], [4.16]). The *potential temperature* is defined "as the temperature a fluid element would attain if it were brought to the surface of the fluid adiabatically and without exchange of salt with the environment" [4.9, p. 55].

To understand the CORIOLIS *force* mentioned several times earlier, we first consider the motion of a mass point $m$ in an accelerated frame of reference (like the surface of the rotating Earth). The angular velocity vector $\vec{\omega}$ of the Earth is parallel to its axis of rotation and has the rate

$$\omega = \frac{2\pi}{24.3600} = 0.72722 \cdot 10^{-4}\,\mathrm{s}^{-1}. \tag{4.2.1}$$

Furthermore

$$\omega^2 R = 3.388\,\mathrm{cm\,s}^{-2}, \tag{4.2.2}$$

where $R$ is the radius of the Earth. In a rotating coordinate system, a time dependent vector $\vec{r}(t)$ transforms as [4.10], [4.9]

$$\frac{\mathrm{d}\vec{r}}{\mathrm{d}t} = \frac{\partial \vec{r}}{\partial t} + [\vec{\omega} \times \vec{r}]. \tag{4.2.3}$$

Here the first right-hand side term designates the change of $\vec{r}$ within the rotating system and the left-hand side term describes the total change. Here $\vec{r}$ is the radius vector (location vector). Then $\dot{\vec{r}} = \mathrm{d}\vec{r}/\mathrm{d}t$ is the mass point velocity $\vec{v}$ and the acceleration $\vec{a}$ is given by $\dot{\vec{v}}$

$$\begin{aligned} \vec{a} = \dot{\vec{v}} = \frac{\mathrm{d}\vec{v}}{\mathrm{d}t} = [\vec{\omega} \times \vec{v}] + \frac{\partial \vec{v}}{\partial t} = \\ [\vec{\omega} \times [\vec{\omega} \times \vec{r}]] + \left[\vec{\omega} \times \frac{\partial \vec{r}}{\partial t}\right] + \left[\frac{\mathrm{d}\vec{\omega}}{\mathrm{d}t} \times \vec{r}\right] + \left[\vec{\omega} \times \frac{\partial \vec{r}}{\partial t}\right] + \frac{\partial \vec{r}}{\partial t}. \end{aligned} \tag{4.2.4}$$

Here $\partial \vec{v}/\partial t$ is the acceleration $\vec{a}\,'$ within the rotating system and $\mathrm{d}\vec{\omega}/\mathrm{d}t$ is the change of the angular (rotational) velocity $\vec{\omega}$. Summing up, the equation of motion of a mass point $m$ in a rotating system may be written

$$m\vec{a} = m\vec{a}\,' + 2m[\vec{\omega} \times \vec{v}\,'] + m[\vec{\omega} \times [\vec{\omega} \times \vec{r}]]. \tag{4.2.5}$$

Here we assumed that the angular rate $\omega$ is constant and given by (4.2.1). In order to obtain the additional forces in the rotating system we rewrite in the form

$$\underset{\text{rel. accel.}}{m\vec{a}\,'} = \underset{\text{abs. accel.}}{m\vec{a}} - \underset{\text{CORIOLIS force}}{2m[\vec{\omega} \times \vec{v}\,']} - \underset{\text{centrifugal force.}}{m[\vec{\omega} \times [\vec{\omega} \times \vec{r}]]}, \tag{4.2.6}$$

The cross product $[\vec{\omega} \times \vec{r}]$ on the surface of the Earth is given by

$$[\vec{\omega} \times \vec{r}] = |\vec{\omega}| \cdot |\vec{r}| \sin(\angle \vec{\omega}, \vec{r}) = \omega R \cos \Phi, \tag{4.2.7}$$

where $\Phi$ is again the geographical latitude. Then the CORIOLIS *acceleration* is defined by

$$\vec{g}' = -2\vec{\omega} \times \vec{v}, \quad f = 2\omega \sin \Phi \quad \text{and} \quad f^* = 2\omega \cos \Phi. \tag{4.2.8}$$

$f$ is the CORIOLIS *parameter* ($= 1.03 \cdot 10^{-4}\,\mathrm{s}^{-1}$ for $\Phi = 45°$). At the equator the CORIOLIS parameter $f$ vanishes. A table for $f(\Phi)$ may be found on p 2-131 in [2.2].

The centrifugal acceleration is given by

$$[\vec{\omega} \times [\vec{\omega} \times \vec{r}]] = \omega^2 R \cos \Phi, \tag{4.2.9}$$

so that

$$g \to g \left(1 - \frac{\omega^2}{g} R \cos^2 \Phi \right) \approx g \left(1 - \frac{1}{289} \cos^2 \Phi \right)$$

expresses the gravity reduction due to the centrifugal force on the surface of the Earth. Here it has been assumed, that the Earth is an exact sphere. In fact it is a geoid and the number 289 should be replaced by 191. Some authors [4.9] define an apparent gravity $\vec{g} = -\nabla \psi$ or a *geopotential* by

$$\psi = \psi_0 - \frac{1}{2} \omega^2 R^2 \cos^2 \Phi. \tag{4.2.10}$$

Now we are able to write down the fluid equations of motion in spherical coordinates [4.9], [4.4] [4.11]. Taking into account the definitions for the three Cartesian components

$$u = r \cos \Phi \frac{\mathrm{d}\lambda}{\mathrm{d}t}, \quad v = r \frac{\mathrm{d}\Phi}{\mathrm{d}t}, \quad w = \frac{\mathrm{d}r}{\mathrm{d}t}, \tag{4.2.11}$$

the EULER *fluid equations* then read on the Earth's surface

$$\begin{aligned}
\rho \left( \frac{\mathrm{d}u}{\mathrm{d}t} - \frac{uv}{R} \tan \Phi + \frac{uw}{R} - fv + f^* w \right) &= -\frac{1}{R \cos \Phi} \frac{\partial p}{\partial \lambda}, \\
\rho \left( \frac{\mathrm{d}v}{\mathrm{d}t} + \frac{u^2}{R} \tan \Phi + \frac{vw}{R} + fu \right) &= -\frac{1}{R} \frac{\partial p}{\partial \Phi}, \\
\rho \left( \frac{\mathrm{d}w}{\mathrm{d}t} - \frac{u^2 + v^2}{R} - f^* u \right) &= -\frac{\partial p}{\partial r} - g\rho.
\end{aligned} \tag{4.2.12}$$

$\Phi$ is the geographical latitude and $\lambda$ is the longitude. Viscosity has been neglected. The continuity equation is given by $\partial \rho / \partial t + \operatorname{div}(\rho \vec{v}) = 0$, $\vec{v} = u, v, w$ or

$$\frac{\partial \rho}{\partial t} + \frac{1}{R} \frac{\partial (\rho v)}{\partial \Phi} + \frac{1}{R \sin \Phi} \frac{\partial (\rho u)}{\partial \lambda} + \frac{\rho v}{R} \operatorname{cotg} \Phi = 0.$$

The CORIOLIS force has two important consequences:

1. it increases the circulation $\Gamma$,
2. it induces the BUYS-BALLOT *theorem* on the rotation of cyclones.

According to (2.6.9) circulation generated by the CORIOLIS force is defined by

$$\frac{\mathrm{d}\Gamma}{\mathrm{d}t} = \oint \vec{g}'\mathrm{d}\vec{s} = 2\oint [\vec{v} \times \vec{\omega}]\mathrm{d}\vec{s} = -2\oint \vec{\omega}[\vec{v} \times \mathrm{d}\vec{s}]. \tag{4.2.13}$$

Here we used the formula for the scalar triple product [1.1]

$$[\vec{A} \times \vec{B}] \cdot \vec{C} = -\vec{B}[\vec{A} \times \vec{C}]. \tag{4.2.14}$$

Thus

$$\begin{aligned} \frac{\mathrm{d}\Gamma}{\mathrm{d}t} &= -\frac{2\mathrm{d}}{\mathrm{d}t}\oint \vec{\omega}[\vec{v}\mathrm{d}t \times \mathrm{d}\vec{s}] = -2\frac{\mathrm{d}}{\mathrm{d}t}\int \vec{\omega}\mathrm{d}\vec{f}, \\ [\vec{v}\mathrm{d}t \times \mathrm{d}\vec{s}] &= \mathrm{d}\vec{f}, \quad F = \int \mathrm{d}\vec{f}. \end{aligned} \tag{4.2.15}$$

Equation (4.2.15) has the following physical consequences:

1. If an air torus (air annulus) migrates towards north (increasing latitude), the area $F$ diminishes and circulation towards east increases,
2. if air rises ($r$ increases) $F$ diminishes and anticlockwise circulation (cyclonic circulation) increases.

The BUYS-BALLOT *theorem* is another consequence of the CORIOLIS force: wind blowing from an aera of higher pressure towards a lower pressure region will be deflected to the right and imposes a cyclonic (counterclockwise) rotation on the northern hemisphere (clockwise on the southern hemisphere).

When air moves without friction or acceleration there is a balance between the CORIOLIS force and the pressure gradient. Such a wind is called *geostrophic wind.*

Several models have been developed to track the path of hurricanes like CLIPER, BAM, NHC90, VICBAR, MRF etc [4.13]. The influence of friction on cyclones have been studied very early [4.14]. It has been found that the *eye* is strongly determined by friction. According to the GULDBERG-MOHN *theorem* [4.15] the inland friction force

$$F \sim 10^{-4}\, v \qquad \mathrm{dyn\,g^{-1}}, \tag{4.2.16}$$

where $v$ is measured in $\mathrm{cm\,s^{-1}}$, is capable to reduce the vorticity of a cyclone to one tenth of its actual vorticity during 17 minutes.

## Problems

1. Derive the equations of motion for a mass point in a spherical coordinate system $r$ latitude $\Phi$ and the longitude $\lambda$, which rotates with the Earth.

   Solution [4.12]:

$$\begin{aligned} &\ddot{r} - r\dot{\Phi}^2 - r\cos^2\Phi\dot{\lambda}(\dot{\lambda} - 2\omega) + g = 0, \\ &r\ddot{\Phi} + 2\dot{r}\dot{\Phi} + r\sin\Phi\cos\Phi\dot{\lambda}(\dot{\lambda} - 2\omega) = 0, \\ &\ddot{\lambda}r\cos\Phi + 2\left(\dot{r}\cos\Phi - r\dot{\Phi}\sin\Phi\right)(\dot{\lambda} - \omega) = 0. \end{aligned} \tag{4.2.17}$$

2. Derive the equations (4.2.11)–(4.2.12), [4.9], [4.11].

3. Discuss the extension of the equations (4.2.11)–(4.2.12) as given on p 59 in [4.9]. They are:

   the equation of continuity:
   $(\vec{v} = u, v, w,\ u = R\cos\Phi\cdot\lambda_t,\ v = R\Phi_t,\ w = z_t)$

$$\rho_t + \mathrm{div}(\rho\vec{v}) = 0, \tag{4.2.18}$$

   the equations of motion

$$\begin{aligned} \frac{1}{\rho}\frac{\partial\rho u}{\partial t} &= -\frac{1}{\rho}\mathrm{div}(\rho u\vec{v}) + \frac{\tan\Phi}{R}uv - \frac{uw}{R} + fv - f^*w - \frac{1}{\rho}\frac{1}{R\cos\Phi}\frac{\partial p}{\partial\lambda} + F_\lambda, \\ \frac{1}{\rho}\frac{\partial\rho v}{\partial t} &= -\frac{1}{\rho}\mathrm{div}(\rho v\vec{v}) - \frac{\tan\Phi}{R}u^2 - \frac{vw}{R} - fu - \frac{1}{\rho}\frac{1}{R}\frac{\partial p}{\partial\Phi} + F_\Phi, \\ \frac{1}{\rho}\frac{\partial\rho w}{\partial t} &= -\frac{1}{\rho}\mathrm{div}(\rho w\vec{v}) + \frac{u^2}{R} + \frac{v^2}{R} + f^*u - \frac{1}{\rho}\frac{\partial p}{\partial z} - g + F_z, \end{aligned} \tag{4.2.19}$$

   where $\vec{F}$ describes friction forces which can be expressed by the stress tensor (2.3.12). The energy theorem reads

$$\frac{1}{\rho}\frac{\partial\rho T}{\partial t} = -\frac{1}{\rho}\mathrm{div}(\rho T\vec{v}) + Q/c_p + \frac{\bar{\lambda}T}{\rho}\frac{\mathrm{d}p}{\mathrm{d}t}, \tag{4.2.20}$$

where $Q$ equals the net heating rate per unit mass (solar radiation, latent heating, frictional heating etc). $\bar{\lambda}$ is again the thermal conductivity. The balance equation for water vapor reads

$$\frac{\mathrm{d}q}{\mathrm{d}t} = s(q) + D, \tag{4.2.21}$$

where $q$ is the specific humidity, $s(q)$ represents sources and sinks of water vapor, compare [2.5], and $D$ describes turbulent eddy diffusion of water vapor into the volume. $s(q)$ can be represented by the difference between the rate of evaporation plus sublimation and the rate of condensation per unit mass. The balance may be written

$$\frac{1}{\rho}\frac{\partial \rho q}{\partial t} = -\frac{1}{\rho}\mathrm{div}(\rho q \vec{v}) + s + D. \tag{4.2.22}$$

The equation of state for moist air is given in the form

$$p = \rho R T(1 + 0.61\, q). \tag{4.2.23}$$

The factor 0.61 is the result of a series expansion on p 53 and 275 in [4.9]. Compare (4.2.19) with (4.2.12).

4. Derive the one-dimensional equations of continuity and of motion in circular cylindrical coordinates [2.6], [1.1].

   Hints: use (2.5.44), (2.9.3) and (2.3.26) to (2.3.30) with $\vec{F} = 0$, $\eta = 0$, $\eta' = 0$, $\kappa$ for air from Table 2.2.

   Solution:

$$\frac{\partial \rho}{\partial t} + \frac{1}{r}\rho u + \frac{\partial \rho}{\partial r}u + \rho\frac{\partial u}{\partial r} = 0, \tag{4.2.24}$$

$$\rho\frac{\partial u}{\partial t} + \rho u\frac{\partial u}{\partial r} + \frac{\partial p}{\partial r} = 0, \tag{4.2.25}$$

$$\frac{\partial p}{\partial r} = \rho^{n-1}a^2\frac{\partial \rho}{\partial r} \quad \text{or} \quad \frac{\partial p}{\partial r} = \kappa\rho^{\kappa-1}\,\text{const}. \tag{4.2.26}$$

   Solve the system by a similarity transformation $\rho = R(\eta)$, $u = U(\eta)$, $p = \kappa\,\mathrm{const}\,R^{\kappa-1}$; $\eta = r^{\alpha}t^{\beta}$.

   Hint: multiply the resulting two equations by $r$.

   Solution: $\alpha = 1$, $\beta = -1$;

$$\begin{aligned} -R'\eta^2 + RU + R'\eta U + RU'\eta &= 0, \\ -RU'\eta + RUU' + \kappa\cdot\mathrm{const}\cdot R^{\kappa-1} &= 0. \end{aligned} \tag{4.2.27}$$

## 4.3 Mathematical modelling of cyclones

Since cyclones and hurricanes are quite complex phenomena involving exterior forces, friction, thermodynamic processes etc it is very difficult to find appropriate mathematical models. Such numerical models have been proposed recently, but we restrict ourselves to some remarks on similarity variables and exercises of the Mathematica code. If appropriate equations are given, the question arises, how to solve them. Several methods may be used like similarity solutions, characteristics, a mass-variable transformation etc and finally numerical methods like *finite elements* and other methods like the GALERKIN method, [1.1, p. 216] or finite differences [4.17], which are however outside the scope of this book. Finally, the equations may be written in various coordinate systems: Cartesian, cylindrical, spherical etc.

Two situations seem to be of interest:

1. investigation of a locally fixed cyclone in circular cylindrical or Cartesian coordinates,
2. the tracking of cyclones over the surface of the Earth.

We will start with a simple cylindrical model. We first choose the coordinates $t, r$ and $\Phi$ and assume independence on $z$. Here $\Phi$ is not the latitude on Earth but a circular cylinder coordinate. We include a pseudo-COROLIS term. The equation of continuity should be

$$\frac{\partial \rho}{\partial t} + \frac{1}{r}\frac{\partial}{\partial r}(r\rho u) + \frac{1}{r}\frac{\partial}{\partial \Phi}(\rho v) = 0 \tag{4.3.1}$$

and the equations of motion are assumed to be [4.12]

$$\rho\left(\frac{\partial u}{\partial t} + u\frac{\partial u}{\partial r} + \frac{v}{r}\frac{\partial u}{\partial \Phi} - \frac{v^2}{r} + fv\right) + \frac{\partial p}{\partial r} = 0, \tag{4.3.2}$$

$$\rho\left(\frac{\partial v}{\partial t} + u\frac{\partial v}{\partial r} + \frac{v}{r}\frac{\partial v}{\partial \Phi} + \frac{vu}{r} - fu\right) + \frac{1}{r}\frac{\partial p}{\partial \Phi} = 0, \tag{4.3.3}$$

where $f$ is the CORIOLIS parameter.

A similarity setup $\zeta = r^{\alpha}t^{\beta}$

$$\begin{gathered} \rho(r,\Phi) = R(\zeta), \quad u(r,\Phi) = U(\zeta), \quad v(r,\Phi) = V(\zeta), \\ p(r,\Phi) = \text{const} \cdot \kappa\rho^{\kappa-1} = cR^{\kappa-1} \end{gathered} \tag{4.3.4}$$

yields satisfaction of (4.3.1), but problems arise even with the pressureless equations of motion. This may be demonstrated by the following Mathematica code

```
(* Cyl simil Coriolis *)
Clear[p,Rt,Rr,Ur,Rf,Vf,Ut,Uf,Vt,a,b,Ko,Kof];
p[r_,t_]=r^a*t^b;(*a=1;b=-1;*)
Rt=D[R[p,fi],p]*D[p[r,t],t];
Rr=D[R[p,fi],p]*D[p[r,t],r];
Ur=D[U[p,fi],p]*D[p[r,t],r];
Rf=D[R[p,fi],fi];
Vf=D[V[p,fi],fi];
Ut=D[U[p,fi],p]*D[p[r,t],t];                                    (4.3.5)
Uf=D[U[p,fi],fi];
Vt=D[V[p,fi],p]*D[p[r,t],t];
Ko=(Rt+R*U/r+Rr*U+Ur*R+Rf*V/r+R*Vf/r)*r;
(* Ko = Continuity equation *)
Clear[Kof,p];Kof=R[p,fi]*U[p,fi]+V[p,fi]*D[R[p,fi],fi]+
R[p,fi]*D[V[p,fi],fi]+D[R[p,fi],p]*
(p[r,t]^2+U[p,fi]*p[r,t])+p[r,t]*R[p,fi]*D[U[p,fi],p]
```

$$\mathrm{R[p,\ fi]\ U[p,\ fi] + V[p,\ fi]\ R^{(0,1)}\ [p,\ fi] + R[p,\ fi]\ V^{(0,1)}\ [p,\ fi] +} \\ \mathrm{(p[r,\ t]^2 + p[r,\ t]\ U[p,\ fi])\ R^{(1,0)}\ [p,\ fi] + p[r,\ t]\ R[p,\ fi] U^{(1,0)}\ [p,\ fi]}$$

```
Clear[M1,p];
(* M1 = Equation of motion (4.3.2) without pressure *)
M1=(R[p,fi]*(Ut+U[p,fi]*Ur+V[p,fi]*Uf/r-V[p,fi]^2/r)+
f*V[p,fi])
```

$$\mathrm{f\,V[p,\ fi] + R[p,\ fi]} \left( -\frac{\mathrm{V[p,\ fi]^2}}{\mathrm{r}} + \frac{\mathrm{V[p,\ fi]\ U^{(0,1)}\ [p,\ fi]}}{\mathrm{r}} + \right. \\ \left. \mathrm{b\ r^a\ t\ t^{-1+b}\ U^{(1,0)}\ [p,\ fi] + a\ r^{-1+a}\ t^b\ U[p,\ fi]\ U^{(1,0)}\ [p,\ fi]} \right)$$

One finds that no combination of $rt^0$, $r^0t^{-1}$ and $r^{-1}t^0$ can be reduced to a power of $p = r/t$. An analogous result can be obtained from (4.3.1) in Cartesian coordinates $p = x^a t^b$, $\rho(x, y, t) = R(p, y)$ etc. Characteristics methods will not be used, since only two independent variables $t, r$ or $t, x$ are not sufficient to describe a cyclone.

As next step we investigate a model in spherical coordinates as described by (4.2.12). Since it seems to be easier to satisfy the continuity equation by a similarity solution than an equation of motion, we investigate the possibility to solve the EULER equation for the $u$-component of (4.2.12). Since we want to investigate the tracking of a cyclone over the surface of Earth we assume $\partial\ /\partial r\ = 0$ and we replace $r$ by the radius R of the Earth. The three equations (4.2.12) may be found in the literature in two different versions.

In [4.9] the pressure term in the $w$-component equation reads $\partial p/\partial z$, but in [4.11] one finds $\partial p/\partial r$. When comparing equations given in spherical coordinates by several authors, one has to be careful: some authors define the equator of the sphere by the geographical latitude $\Phi =$ or $\vartheta = 0°$, but others use $\Phi = 90°$. The geographical longitude is denoted by $\lambda$ or $\Phi$. The $u$-component is identical with $v_\Phi$, $v \to v_\vartheta$, $w \to v_r$ and $\cos\Phi$ becomes $\sin\vartheta$ and $\tan\Phi$ is replaced by $\cot\vartheta$.

We now consider the $u$-component of (4.2.12) in the form

$$\frac{\partial u}{\partial t} + \frac{u}{R\cos\Phi}\frac{\partial u}{\partial\lambda} + \frac{v}{R}\frac{\partial u}{\partial\Phi} - \frac{uv}{R}\tan\Phi + \frac{uw}{R} - v2\omega\sin\Phi + \\ w2\omega\cos\Phi + \frac{1}{R}\frac{\partial P}{\partial\lambda} = 0. \tag{4.3.6}$$

Here we used (2.3.17) to get rid of the density $\rho$. $P$ is again the *pressure-density integral.* We used the setup

$$\zeta(t,\lambda) = t^a\lambda^b \tag{4.3.7}$$

which eliminates one of the three independent variables $t, \lambda, \Phi$ of (4.3.6). Furthermore

$$u(t,\lambda,\Phi) = U(\zeta(t,\lambda),\Phi)t^c\lambda^d, \qquad v(t,\lambda,\Phi) = V(\zeta(t,\lambda),\Phi)t^e\lambda^f, \\ w(t,\lambda,\Phi) = W(\zeta(t,\lambda),\Phi), \qquad p \to P(\zeta(t,\lambda),\Phi)\lambda^g. \tag{4.3.8}$$

Then we used the following definitions in the Mathematica code:

$$ut = \frac{\partial u}{\partial t}, \quad ul = \frac{\partial u}{\partial\lambda}, \quad uf = \frac{\partial u}{\partial\rho}, \quad pl = \frac{\partial P}{\partial\lambda}. \tag{4.3.9}$$

Inserting into (4.3.6) immediately yields $g = 1$. Then the code

```
(* sphericalsim.nb  Spherical coordinates and
similarity transformations *)
Clear[ζ,a,b,c,d,e,f,g,U,V,W,u,v,w,ut,ul,uf,pl];
ζ[t_,λ_]=t^a*λ^b;d=1;
u[ζ[t,λ],Φ]=U[ζ[t,λ ],Φ]*t^c*λ^d;
v[ζ[t,λ],Φ]=V[ζ[t,λ ],Φ]*t^e*λ^f;
w[ζ[t,λ],Φ]=W[ζ[t,λ ],Φ];
ut=D[U[ζ[t,λ],Φ]*t^c*λ^d,t];
ul=D[U[ζ[t,λ],Φ]*t^c*λ ^d,λ];
uf=D[U[ζ[t,λ],Φ]*t^c*λ ^d,Φ];
```

```
pl=D[P[ζ[t,λ],Φ],λ];
ut+u[ζ[t,λ],Φ]*ul/(R*Cos[Φ])+v[ζ [t,λ],Φ]*uf/R-
u[ζ[t,λ],Φ]*v[ζ[t,λ ],Φ]*Tan[Φ]/R+
u[ζ[t,λ],Φ]*w[ζ[t,λ ],Φ]/R-v[ζ[t,λ],Φ]*2*ω *Sin[Φ]+
w[ζ[t,λ],Φ]*2*ω*Cos[Φ ]+pl/(R*Cos[Φ])
```

$$
\begin{aligned}
& \mathrm{c}\ \mathrm{t}^{-1+\mathrm{c}}\ \lambda\ \mathrm{U}[\mathrm{t}^{a}\ \lambda^{\mathrm{b}},\ \Phi]\ \text{-}\ 2\ \mathrm{t}^{\mathrm{e}}\ \lambda^{\mathrm{f}}\ \omega\ \mathrm{Sin}[\Phi]\ \mathrm{V}[\ \mathrm{t}^{\mathrm{a}}\ \lambda^{\mathrm{b}},\ \Phi]\ \text{-} \\
& \frac{\mathrm{t}^{\mathrm{c+e}}\ \lambda^{1+\mathrm{f}}\ \mathrm{Tan}[\Phi]\ \mathrm{U}[\mathrm{t}^{\mathrm{a}}\ \lambda^{\mathrm{b}},\Phi]\ \mathrm{V}[\mathrm{t}^{\mathrm{a}}\ \lambda^{\mathrm{b}},\ \Phi]}{\mathrm{R}} + 2\ \omega\ \mathrm{Cos}[\Phi]\ \mathrm{W}[\mathrm{t}^{\mathrm{a}}\ \lambda^{\mathrm{b}},\Phi] + \\
& \frac{\mathrm{t}^{\mathrm{c}}\ \lambda\ \mathrm{U}[\mathrm{t}^{\mathrm{a}}\ \lambda^{\mathrm{b}},\ \Phi]\ \mathrm{W}[\mathrm{t}^{\mathrm{a}}\ \lambda^{\mathrm{b}},\Phi]}{\mathrm{R}} + \frac{\mathrm{t}^{\mathrm{c+e}}\ \lambda^{1+\mathrm{f}}\ \mathrm{V}[\mathrm{t}^{\mathrm{a}}\ \lambda^{\mathrm{b}},\ \Phi]\ \mathrm{U}^{(0,1)}\ \mathrm{t}^{\mathrm{a}}\ \lambda^{\mathrm{b}},\Phi\ ]}{\mathrm{R}} + \\
& \frac{\mathrm{b}\ \mathrm{t}^{\mathrm{a}}\ \lambda^{-1+\mathrm{b}}\ \mathrm{Sec}[\Phi]\ \mathrm{P}^{(1,0)}\ [\mathrm{t}^{\mathrm{a}}\ \lambda^{\mathrm{b}},\ \Phi]}{\mathrm{R}} + \mathrm{a}\ \mathrm{t}^{-1+\mathrm{a}+\mathrm{c}}\ \lambda^{1+\mathrm{b}}\ \mathrm{U}^{(1,0)}\ [\mathrm{t}^{\mathrm{a}}\ \lambda^{\mathrm{b}},\ \Phi] + \\
& \frac{\mathrm{t}^{\mathrm{c}}\ \lambda\ \mathrm{Sec}[\Phi]\ \mathrm{U}[\mathrm{t}^{\mathrm{a}}\ \lambda^{\mathrm{b}},\ \Phi]\ (\mathrm{t}^{\mathrm{c}}\ \mathrm{U}[\mathrm{t}^{\mathrm{a}}\ \lambda^{\mathrm{b}},\ \Phi]\ +\ \mathrm{b}\ \mathrm{t}^{\mathrm{a+c}}\ \lambda^{\mathrm{b}}\ \mathrm{U}^{(1,0)}\ [\mathrm{t}^{\mathrm{a}}\ \lambda^{\mathrm{b}},\ \Phi])}{\mathrm{R}}
\end{aligned}
\tag{4.3.10}
$$

demonstrates that not all terms cancel. Collecting $t^a\lambda^b$ into $\zeta$, we have additionally the terms

$$
t^{c-1}, \quad t^e\lambda^f, \quad t^{c+e}\lambda^{1+f}, \quad t^c\lambda, \quad \lambda^{-1}, \quad t^{2c}\lambda, \quad t^{2c}, \tag{4.3.11}
$$

if we choose $d = 1$. In order that these terms cancel, one has to assume

$$
\begin{array}{lccccccccc}
t: & c-1 & = & e & = & c+e & = & c & = & 2c, \\
\lambda: & 1 & = & f & = & 1+f & = & -1, & &
\end{array} \tag{4.3.12}
$$

which gives contradictions. We thus have to conclude that a similarity solution of (4.3.6) can not be found. Apparently only numerical methods can solve the system (4.2.12).

## Problems

1. Write down the equations of continuity and of motion in spherical coordinates using symmetry $\partial/\partial\Phi = 0$, $\vec{v}_\Phi = 0$. (Subscripts designate components and not derivatives.)

$$
\frac{\partial\rho}{\partial t} + \frac{1}{r^2}\frac{\partial}{\partial r}\left(r^2\rho v_r\right) + \frac{1}{r\sin\vartheta}\frac{\partial}{\partial\vartheta}\left(\sin\vartheta\rho v_\vartheta\right) = 0, \tag{4.3.13}
$$

radial momentum:

$$\rho\left(\frac{\partial v_r}{\partial t}+v_r\frac{\partial v_r}{\partial r}+\frac{v_\vartheta}{r}\frac{\partial v_r}{\partial \vartheta}-\frac{v_\vartheta^2}{r}\right)+\frac{\partial p}{\partial r}=0, \tag{4.3.14}$$

azimuthal momentum:

$$\rho\left(\frac{\partial v_\vartheta}{\partial t}+v_r\frac{\partial v_\vartheta}{\partial r}+\frac{v_\vartheta}{r}\frac{\partial v_\vartheta}{\partial \vartheta}+\frac{v_\vartheta v_r}{r}\right)+\frac{1}{r}\frac{\partial p}{\partial \vartheta}=0 \tag{4.3.15}$$

and apply a similarity transformation [1.1]

$$\eta = rt^{\alpha}, \quad \vartheta=\bar{\vartheta}; \quad \rho(r,t,\vartheta)=\bar{\rho}(\eta,\vartheta), \quad p(r,t,\vartheta)=\bar{p}(\eta,\vartheta)t^{-2\alpha},$$
$$v_r(r,t,\vartheta)=\bar{v}_r(\eta)t^{-\alpha}, \quad v_\vartheta(r,t,\vartheta)=\bar{v}_\vartheta(\eta)t^{-\alpha} \tag{4.3.16}$$

and transform these three equations into three partial differential equations for $\bar{\rho}$, $\bar{p}$, $\bar{v}_r$, $\bar{v}_\vartheta$ depending only on $\eta$ and $\vartheta$. The three independent variables $r, t, \vartheta$ have been reduced to two: $\eta$, $\vartheta$.

Solution: $\alpha = 1/2$, $\eta = rt^{-1/2}$ and for the continuity equation the solution is

$$-\frac{\eta}{2}\frac{\partial \bar{\rho}}{\partial \eta}+\frac{2}{\eta}\bar{\rho}\bar{v}_r+\frac{\partial}{\partial \eta}(\bar{\rho}\bar{v}_r)+\frac{\operatorname{cotg}\vartheta}{\eta}\bar{v}_\vartheta\bar{\rho}+\frac{1}{\eta}\frac{\partial}{\partial\vartheta}(\bar{\rho}\bar{v}_\vartheta)=0. \tag{4.3.17}$$

2. After the elimination of $r$ and $t$ due to $\eta$, you may eliminate $\eta$ or $\vartheta$ from the partial differential equations for $\bar{\rho}(\eta,\vartheta)$, $\bar{v}_r(\eta,\vartheta)$, $\bar{v}_\vartheta(\eta,\vartheta)$. Also $p$ may be expressed by $\rho$ due to the adiabatic law $p = \text{const}\cdot\rho^{\kappa}$. Ordinary differential equations may then be integrated by numerical methods like NDSolve with Mathematica.

3. Solve the two equations (4.2.24) and (4.2.25) in cylindrical coordinates by a similarity transformation using $\vec{F}=0$, $\eta=0$, $\eta'=0$, see (2.3.26) to (2.3.30). $\rho = R(r,t) = R(\zeta)$, $\zeta = r^{\alpha}t^{\beta}$, $u(r,t) = U(\zeta)$, $p(r,t) = (\rho^n/n)a^2$, see (2.4.3), (4.2.26), $n=\kappa=1.405$ (Table 2.2).

   Solution:
$$\begin{aligned}&-R'\zeta^2+RU+R'\zeta U+RU'\zeta=0,\\&-RU'\zeta+RUU'+R^{\kappa-1}a^2R'=0,\end{aligned} \tag{4.3.18}$$

   compare with (4.2.27). A Mathematica code to solve these two ordinary equations could be (4.3.19) ($\zeta$ replaced by $z$, $\kappa$ by $n$). Give any numerical value for $a$.

```
Eq1=-R'[z]*z^2+R[z]*U[z]+R'[z]*U[z]*z+R[z]*U'[z]*z;
Eq2=-R[z]*U'[z]*z+R[z]*U[z]*U'[z]+R^(n-1)*a^2*R'[z];
NDSolve[{Eq1 == 0, Eq2 == 0, R[0] == 1.,
U[0] == 1.},{R[z],U[z]},{z,10.}]
```
(4.3.19)

4. Transform the cylindrical coordinates equations (4.2.24), (4.2.25) using the mass-variable transformation.

   Hint:

$$m(r,t) = \int \rho(r,t) 2\pi r \mathrm{d}r,$$
$$\frac{\partial m}{\partial r} = \rho 2\pi r, \quad \frac{\partial m}{\partial t} = -\rho 2\pi r \hat{u}, \quad \frac{\mathrm{d}r}{\mathrm{d}t} = \hat{u}. \tag{4.3.20}$$

   Apply the transformation $\hat{\rho} = 1/\hat{s}$ and find that now the continuity equation becomes a nonlinear equation

$$-s_t + \hat{u}_m 2\pi r + \frac{1}{r}\frac{\hat{u}}{\hat{s}} = 0, \tag{4.3.21}$$

   compare with (3.6.12)!

## *4.4 Multifluid cyclone modelling*

A *plasma* is an ionized gas consisting of electrons, ions and neutral particles (atoms, molecules). It is described by a *multifluid theory* [2.8], [3.22]. Some plasma properties may be found in strong electrolytes or even in semiconductors. A cyclone which contains liquid water condensate, dry air and water vapor should also be described by a multifluid theory [4.18], [2.5] (1947) or by a two-fluid model [4.4].

The *eye* wall separates an inner region of low humidity and the saturated air outside. An interface between two fluids of different densities or streaming with different speeds is unstable. A slight perturbation like $\exp(ikx - i\omega t)$ of speed or pressure initiates a growing instability. A speed jump $v$ between two adjacent fluid layers gives rise to the KELVIN-HELMHOLTZ *instability* (*velocity shear instability*) and the RAYLEIGH-TAYLOR *instability* is caused by a perturbation of the interface between a heavy fluid ($\bar{\bar{\rho}}$) supported by a light fluid ($\bar{\rho}$) in a gravitational field in $z$-direction [2.8], [3.22], [4.14].

The investigation if a flow process is stable should follow the following scheme [2.10]: the unperturbed flow defined by $\partial\ /\partial t = 0$ and $\vec{v}_0, p_0$ will be superposed by small perturbations $\vec{v}_1, p_1$, where $\vec{v}_1^2 \approx 0$ will be assumed.

The unperturbed steady situation may be defined by

$$\rho_0(\vec{v}_0\nabla)\vec{v}_0 = -\nabla p_0, \quad \operatorname{div}\vec{v}_0 = 0 \tag{4.4.1}$$

and the time dependent perturbations are described by

$$\rho\left(\frac{\partial \vec{v}_1}{\partial t} + (\vec{v}_0\nabla)\,\vec{v}_1\right) = -\nabla p_1, \quad \operatorname{div}\vec{v}_1 = 0. \tag{4.4.2}$$

Applying the operator div on the first equation, one obtains

$$\operatorname{div}\nabla p_1 = \Delta p_1 = 0. \tag{4.4.3}$$

Assuming that the local displacement $\eta(x,t)$ of the interface normal to the $z$-direction is small, then its speed is for constant $x$ given by

$$\frac{\partial \eta}{\partial t} = v_{1z(\text{interface})} - v_{0x}\frac{\partial \eta}{\partial x}. \tag{4.4.4}$$

Assuming

$$p_1 = f(z)\exp(ikx - i\omega t) \tag{4.4.5}$$

and inserting into (4.4.3) one obtains

$$\frac{\mathrm{d}^2 f}{\mathrm{d}z^2} - k^2 f = 0, \quad f(z) = \text{const}\ \exp(\pm kz). \tag{4.4.6}$$

Then (4.4.2) yields

$$v_{1z} = \frac{k\bar{p}_1}{i\bar{\rho}(kv_{0x} - \omega)}. \tag{4.4.7}$$

The bar $\bar{p}_1$ indicates the side of the interface defined by $z > 0$ and two bars $\bar{\bar{p}}_1$ define quantities on the side $z < 0$. Assuming $\eta \sim \exp(ikx - i\omega t)$ we obtain from (4.4.4) the result

$$v_{1z} = \text{const}\ \exp(ikx - i\omega t) = i\eta(kv_{0x} - \omega), \tag{4.4.8}$$

so that

$$\bar{p}_1 = -\frac{\eta\bar{\rho}(kv_{0x} - \omega)^2}{k}. \tag{4.4.9}$$

The assumption $v_{0x} = 0$ for $z < 0$ (fluid at rest) yields

$$\bar{\bar{p}}_1 = \frac{\eta\bar{\bar{\rho}}\omega^2}{k}. \tag{4.4.10}$$

Due to $\bar{p}_1 = \bar{\bar{p}}_1$ one has $\bar{\rho}(kv_{0x} - \omega)^2 = -\bar{\bar{\rho}}\omega^2$, which gives the *dispersion relation*

$$\omega = kv_{0x}\frac{\bar{\rho} \pm i\sqrt{\bar{\rho}\bar{\bar{\rho}}}}{\bar{\rho} + \bar{\bar{\rho}}}. \tag{4.4.11}$$

Since $\omega$ is complex with a positive imaginary part, it indicates the instability of an interface between two regions of different densities and/or different velocities (RAYLEIGHT-TAYLOR resp. KELVIN-HELMHOLTZ-instability). According to [4.14], cyclones are generated by the combined effects of such an instability together with the CORIOLIS force. The discontinuity in $v$ and density is a main factor in cyclone generation [4.4]. Surface transfer of heat from the ocean is another important factor.

Density differences may be due to temperature differences between the warm surface of the ocean and a cold mass of air. The vertical component $\ddot{r}$ in (4.2.17) of the CORIOLIS force may induce a deviation ("tracking of cyclones") to the west for rising masses of air. The problem of tropical cyclone tracking is described by several mathematical models: CLIPER, NOGAPS, MRF, LIBAR etc, see the reports of the National Hurricane Center [4.19]. These model equations have to be solved numerically.

A three-fluid model may be found in [4.11], [4.18]. In this model contributions to the mass flux across the ground surface due to precipitation, evaporation and dew are taken into account. Thus boundary conditions are only satisfied if these effects are considered. In describing mass balances one uses the index 0 for dry air, the index 1 for water vapor and 2 for water droplets. The three components satisfy three equations of continuity:

$$\frac{\partial \rho_k}{\partial t} + \operatorname{div}(\rho_k \vec{v}_k) = D_k, \quad k = 1, 2, 3, \tag{4.4.12}$$

where the local production rates $D_k$ are defined in (2.2.2) [2.5]. Due to the conservation of the total mass, one has

$$\sum_k D_k = 0. \tag{4.4.13}$$

Such considerations may determine net mass transport from the ocean to the atmosphere [4.18], [4.11].

Evaporation is determined by the actual saturation pressure $p_{\text{sat}}$ (2.1.6), it depends on temperature and humidity respectively, see Table 2.1. Precipitation originates in the cold air above about 700 mb, droplets or ice particles melt in the lower warm layer and may freeze upon contact with cold ground.

Saturation pressures calculated according to the MAGNUS *formula* [4.11] (2.6.1) coincide quite well with the values given in Table 2.1: (2.1.6) yields $p_{sat}$ (25° C) = 31.79 mb, whereas the Table gives 31.68 mb. (Remember: 1 kPa=$10^3 \cdot 10^{-5}$ bar = $10^{-2}$ bar = 10 mb; 1 mbar = 100 Pa). *Evaporation* $E$ is determined by (2.4.9) and is usually measured in mm/year or in $\mathrm{kg\,m^{-2}\,s^{-1}}$. It may have a value $2 \cdot 10^{-5}\,\mathrm{kg\,m^{-2}\,s^{-1}}$ [4.18] and may reach an annual-mean evaporation rate up to 2000 mm/year [4.9]. For a hurricane speed $v = 50\,\mathrm{m\,s^{-1}}$, see Table 4.1, one may calculate $E(p_{act}, T)$. For $t = 25°$ C this gives values between 1 to 160 and more mm/year.

## Problems

1. On the line of reasoning of problem 2 in section 2.2, calculate evaporation rates $D$ and humidity for the values of $E$ given above.

2. Calculate $E(p_{act}, T)$ for various values of $v$ (and $T, p_{act}$).

# References


[1.1] Cap, F.: Mathematical Methods in Physics and Engineering. Chapman & Hall/CRC, Boca Raton 2003

[1.2] Morse, P., Feshbach, H.: Methods of Theoretical Physics. McGraw-Hill, New York 1953 (2 volumes)

[1.3] Sommerfeld, A.: Partial Differential Equations in Physics. Academic Press, New York 1949

[1.4] Spanier, J., Oldham, K.: An Atlas of Functions. Springer, Heidelberg 1987

[1.5] Abramowitz, M., Stegun, I.: Handbook of Mathematical Functions. Dover, New York 1972

[1.6] Cap, F.: Amplitude Dispersion and Stability of Nonlinear Weakly Dissipative Waves. *J. Math. Phys.* 13, Nr. 8 (1972), 1126–1130

[1.7] Shivamoggi, B.: Theoretical fluid dynamics. Nijhoff, Dordrecht 1985

[2.1] Cohen, R., Giacomo, P.: Symbols, Units, Nomenclature and Fundamental Constants in Physcis. Document I.U.P.A.P-25, Paris 1987

[2.2] American Institute of Physics Handbook. McGraw-Hill, New York 1957

[2.3] Kuchling, H.: Taschenbuch der Physik. Verlag Deutsch, Frankfurt/M 1986

[2.4] There are many good textbooks on thermodynamics, e.g.
Abbot, M., Van Ness, H.: Thermodynamics. Schaum, McGraw-Hill, New York 1972;
Tolman, R.: The Principles of Statistical Mechanics. Oxford University Press;
Landau, L., Lifschitz, E.: Statistische Physik. Akademie-Verlag, Berlin;
Planck, M.: Vorlesungen über Thermodynamik. de Gruyter, Berlin;
Riedi, P. C.: Thermal Physics. Mac Millan, London 1976

[2.5] Cap, F.: Über eine Erweiterung der Strömungs- und der Kontinuitätsgleichung der instationären Gasdynamik für den Fall des Vorhandenseins von Gasquellen und des Mitgerissenwerdens fester oder flüssiger Partikel. *Acta Phys. Austriaca* 1, Nr. 1 (1947), 89–97

[2.6] Hughes, W., Gaylord, E.: Basic Equations of Engineering Science. Schaum, New York 1962

[2.7] Giles, R.: Fluid Mechanics and Hydraulics. Schaum, New York 1962

[2.8] Cap, F.: Lehrbuch der Plasmaphysik und Magnetohydrodynamik. Springer, Wien 1994

[2.9] Montroll, E.: Encyclopaedic Dictionary of Physics. Thewlis, J. (Ed.). Pergamon Press, Oxford 1961, vol. 2, p. 467

[2.10] Landau, L., Lifschitz, E.: Hydrodynamik. Akademie-Verlag, Berlin 1960

[2.11] Crocco, L.: Eine neue Stromfunktion für die Erforschung der Bewegung der Gase und Rotation. *ZAMM* 17, 1 (1937)

[2.12] Cap, F.: Zum Problem der instationären Stoßpolaren. *Helvetica Physica Acta* 21, Nr. 6 (1948), 505–512;
Cap, F.: A New Equation for the Non-stationary Shock Wave. *J. Chem. Phys.* 17, Nr. 1 (1949), 106–107

[2.13] Ames, W. F.: Nonlinear Partial Differential Equations in Engineering. Academic Press, New York 1965, p. 457

[2.14] Sauer, R.: Elementare Lösungen der Wellengleichung isentropischer Gasströmungen. *ZAMM* 31, Nr. 11 (1951), 339;
Hadamard, I., Leçons sur la propagation des ondes. Paris 1903

[2.15] Bechert, K.: Zur Theorie ebener Störungen in reibungsfreien Gasen. *Ann. Physik* [5], 37, Nr. 2 (1940), 89–123;
Bechert, K.: Zur Theorie ebener Störungen in reibungsfreien Gasen, part 2. *Ann. Physik* [5], 38, Nr. 1 (1940), 1–25;
Darboux, G.: Memoir on partial differential equations of second order, (1870), and *Théorie des surfaces*, Vol. 2, Paris (1915)

[2.16] Riemann, B.: Über die Fortpflanzung ebener Luftwellen von endlicher Schwingungsweite. *Abhandlungen der Gesellschaft der Wissenschaften zu Göttingen*, Mathematischphysikalische Klasse 8, Nr. 43 (1860);
`http://www.emis.de/classics/Riemann[Wilkins]`;
`http://www.mathematik.ch/mathematiker/riemann.php`

[2.17] Bechert, K.: Über die Ausbreitung von Zylinder- und Kugelwellen in reibungsfreien Gasen und Flüssigkeiten. *Ann. Physik* [5], 39 Nr. 3 (1941), 169–202

[2.18] Bechert, K.: Über die Differentialgleichungen der Wellenausbreitung in Gasen. *Ann. Physik* [5], 39 Nr. 3 (1941), 357–372

[2.19] Courant, F., Friedrichs, K.: Supersonic Flow and Shock Waves. Wiley-Interscience, New York 1948

[2.20] Lax, P. D.: *Comm. Pure. Appl. Math.* 10, 537 (1957) according to [2.13]; see also Fox, L.: Numerical Solution of Ordinary and Partial Differential Equations. Pergamon, Oxford 1962, p. 355;
Von Neumann, J., Richtmyer, R.: A Method for the Numerical Calculation of Hydrodynamic Shocks. *J. Appl. Phys.* 21, March 1950, p. 232; see also Lax-Wendroff, *Comm. Pur. Appl. Math.* 13, (1960) 217–237

[2.21] Preiswerk, E.: Anwendungen gasdynamischer Methoden auf Wasserströmungen mit freier Oberfläche. Dissertation 1938, Eidgenössische Technische Hochschule Zürich

[3.1] Sube, R., Eisenreich, G.: Wörterbuch Physik, 3 Vols. Harri Deutsch, Frankfurt/M 1973;
Thewlis, J.: Encyclopaedic Dictionary of Physics with Multilingual Glossary, 9 vols. Pergamon Press, Oxford 1964

[3.2] Titov, V., Gonzales, F.: Implementation and Testing of the Method of Splitting Tsunami (MOST) Model, NOAA Technical Memorandum ERL PMEL-112;
Murty, T.: Storm surges-meteorological ocean tides. Canad. Bull. Fisheries Aquatic Sci. 212, p. 897

[3.3] `http://www.pmel.noaa.gov/tsunami/indo20041 226/;`
`http://www.caets.org/NAE/bridgecom.nsf/weblinks/MKEZ-6DFR`

[3.4] `http://www.walrus.we.usgs.gov/tsunami/basic.html`

[3.5] Bechert, K.: Ebene Wellen in idealen Gasen mit Reibung und Wärmeleitung. *Ann. Physik* [5], 40, Nr. 3 (1941), 207–248

[3.6] Dressler, R.: Mathematical Solution of the Problem of Roll-Waves in Inclined Open Channels. *Comm. Pure Appl. Math.* 2, Nr. 2 (1949), p. 149–194

[3.7] Lowell, S., C.: The Propagation of Waves in Shallow Water. *Comm. Pure App. Math.* 2, Nr. 2 (1949), p. 275–291

[3.8] Review articles in M. Lakshmann (Ed.): Solitons. Introduction and Applications. Springer, Berlin 1988

[3.9] Nettel, S.: Wave Physics, Oscillations-Solitons-Chaos. Springer, Berlin 1992

[3.10] Kadomtsev, B., Karpman, V.: Nonlinear Waves. *Sov. Phys. Uphekhi* 14, 40 (1971), and Cap, F.: Handbook on Plasma Instabilities, Vol 3. Academic Press, New York 1982

[3.11] Jeffrey, A., Taniuti, T.: Non-linear Wave Propagation. Academic Press, New York 1964

[3.12] Sauerwein, H.: Numerical Calculations of Arbitrary Multidimensional and Unsteady Flows By the Method of Characteristics, May 1966, Air Force Report Nr. BSD-TR-66-169 and Aerospace Corp., San Bernardino;
Fox, L.: Characteristics in three independent variables, p. 366–377 in Numerical Solution of Ordinary and Partial Differential Equations, Summer School Oxford August 1961, Pergamon Press, Oxford, 1962;
Butler, D. S.: The numerical solution of hyperbolic systems of differential equations in three independent variables. *Proc. Roy. Soc. A* 255, 232–252;
Alder, B., Fernbach, S., Rotenberg, M.:. *Methods in Computational Physics*, Vol. 3, Fundamental Methods in Hydrodynamics. Academic Press, New York 1964

[3.13] Dias, F., Dutyckh, D.: Dynamics of Tsunami Waves. 2005 Springer, found in the internet
`http://www.cmla.ens-cachan.fr/Utilisateurs/dias`

[3.14] Provis, D., Radok, R. (Eds.): Waves on Water of Variable Depth. *Lecture Notes in Physics*, Springer, Berlin, and Australian Academy of Science, Canberra 1977

[3.15] Ansorge, R.: Mathematical Methods in Fluiddynamics. Wiley-VCH, Weinheim 2003

[3.16] Kirby, J., Wei, G., Chen, Q., Kennedy, A., Dalrymple, R.: Fully Nonlinear Boussinesq Wave Model. Documentation and User's Manual,

FUNWAVE 1.0, Research Report NO.CACR-98-06, September 1998 and *J. Wtrwy, Port, Coast Ocean Engrg.* 120, p. 251–261, see also 126 (2000), p. 39–47;
Bona, J., Chen, M., Saut, Y.: Boussinesq Equations and Other Systems for Small-Amplitude Long Waves in Nonlinear Dispersive Media. *J. of Nonlinear Science* (Springer, New York) 12, Nr. 4, July 2002, p. 283–318;
Gröbner, W.: Oberflächenwellen von Flüssigkeiten (Variational Calculus), *Annali della Scuola Normale Superiore di Pisa* Ser. III, Vol V, Fasc. III–IV, 1951, p. 175–191;
Madsen, P., Bingham, H., Schäffer, H.: Boussinesq-type formulation for fully nonlinear and extremely dispersive water waves: derivation and analysis (Series solutions). *Roy. Soc. Proc. Math. Phys. Eng. Sci.* 459, Nr. 2033, May 2003, p. 1075–1104;
Kennedy, B., Dalrympel, R., Kirby, J., Chen, Q.: Determination of Inverse Depths Using Direct Boussinesq Modeling. *J. Wtrwy, Port, Coast Ocean Engrg.* 126 (2000), p. 206–214;
Nagata, Y.: Survey of theoretical research into tsunamis and observations of actual tsunamis in Java. *Lecture Notes in Physics*, Vol. 64. Springer (1977), p. 49–62;
Wei, G., Kirby, J., Grilli, S., Subramanya, R.: A fully nonlinear Boussinesq model for surface waves I: *J. Fluid Mechs* 294, p. 71–92;
Chen, Q., Dalrymple, R., Kirby, J., Kennedy, A., Haller, M.: Boussinesq modeling of a rip current system. *J. Geophys. Res.* 104 (1999), Nr 20, p. 637; see also [3.15], [3.23]

[3.17] Rzadkiewicz, A., Mariotte, C., Heinrich, P.: Numerical Simulation of Submarine Landslides and their Hydraulic Effects. *J. Wtrwy, Port, Coast Ocean Engrg.* 123 (1997), p. 149–157;
Western Coastal and Marine Geology, USGS (see [3.19]): Tsunami Generation from the 2004 M=9.0 Sumatra Earthquake;
Okada, K.: Surface deformation due to shear and tensile faults in a halfspace. *Bull.Seism.Soc.Am.* 75, 1135–1154;
Lay, T., Kanamori, H. et al.: The Great Sumatra-Andaman Earthquake of 26 December 2004. *Science* 308, 20 May 2005, p. 127–1132 (energy release of earthquake $10^{18}$ J, tsunami $4.2 \cdot 10^{15}$ J)

[3.18] Monaghan, J.: Simulating Free Surface Flows with SPH. *J. of Computational Physics* 110, Nr 2, Feb. 1994, p. 399–406;
Titov, V., Synolakis, C.: Numerical Modeling of Tidal Wave Runup.

*J. Wtrw. Port, Coast Ocean Engrg.* 124, Nr. 4, August 1995, p. 157–171;
Wei, G., Kirby, J.: A Time-Dependent Numerical Code for Extended Boussinesq Equations. *J. Wtrwy, Port, Coast Ocean Engrg.* 120, (1995), 251–261;
Nobuaki, K.: Numerical simulation of tsunami hazard potential. *Natural Hazards* 29, Nr. 3, p. 425–436

[3.19] http://walrus.wr.usgs.gov/tsunami
http://www.pmel.noaa.gov/tsunami (and tsunamihazard)
http://wcatwc.arh.noaa.gov/main.htm (West Coast)
http://www.prh.noaa.gov/pr/ptwc (Pacific)
http://www.tsunamiwave.info
http://www.ngdc.noaa.gov/seg/hazards/tsevsrch_idb.shtml
http://walrus.wr.usgs.gov/tsunami/links.html
http://www.fema.gov/hazards/tsunamis/tsunamif.shtm
http://ask.usgs.gov/questions.html
http://nctr.pmel.noaah.gov
http://www.wakayama_nct.ac.jp
http://www.tsunami.civil.tohoku.ac.jp/gahkasyoukai/kan/staff/koike
http://www.drs.dpri.kyoto-u.ac.jp/sumatra/index-e.html
http://www.tsunami.civil.tohoku.ac.jp/hokusai2/main/eng/koike.html/06.02.2006
http://www.cmla.ens-cachan.fr/Utilisateurs/dias
http://www.en.wikipedia.org/wiki/shock_wave
see also [3.3], [3.4]

[3.20] Karpman, V., (translation Cap, F.): Nonlinear Waves in Dispersive Media. Pergamon Press, Oxford 1975

[3.21] Models in mathematical Physics, unknown author, see
http://www.mathe.tu-freiberg.de/inst/amm1/Mitarbeiter Sproessig, Boussinesq equation
in http://www.google.at a search gives about 189.000 results for Boussinesq equation and a list of 126 references is given in
http://www.comp.leeds.ac.uk/markw/Research/PhD/REFS
The list has been updated 1/4/2003
http://mathworld.wolfram.com/BoussinesqEquation.html
http://scienceworld.wolfram.com/physics

/BoussinesqEquation.html
see also [3.16], [3.18] (Wei and Kirby), or [3.16] Madsen et al.

[3.22] Cap, F.: Handbook on Plasma Instabilities, 3 vols. Academic Press, New York 1982, p. 307; 326; 1134

[3.23] http://www.efm.leeds.ac.uk/CIVE/StVenant.pdf
Meis, T., Marcowitz, U.: Numerische Behandlung partieller Differentialgleichungen. Springer, Berlin 1978;
Gerald, C.: applied numerical analysis, addison-wesley, reading, mass. 1970;
Gruber, R.:. Finite Elements in Physics. North Holland, Amsterdam 1986;
Zhou, J., Causon, D. et al.: Numerical Prediction of Dam-break Flows in General Geometries with Complex Bed Topography. *J. Hydraulic Engrg.* 130, Nr. 4, April 2004, p. 332–340;
Ying, X., Khan, A., et al.: Upwind Conservative Scheme for the Saint Venant Equations. *J. Hydraulic Engrg.* 130, Nr. 10, October 2004, p. 977–987;
Fagherini, S., Rasetarinera, P., et al.: Numerical solution of the dam-break problem with a discontinuous Galerkin method. *J. Hydraulic Engrg.* 130, Nr. 6, 2004, p. 532–539; see also *J. Hydraulic Res.*;
Morel, A., Frey, M., et al.: Multidimensional high order method of transport for the shallow water equations. Zürich, Eidgen. Tech. Hochschule 1996 (Seminar angewandte Mathematik);
Löhner, R.: Applied computational fluid dynamics techniques – an introduction based on finite element methods. Wiley, Chichester 2001;
Chung, T.: Computational fluid dynamics. Cambridge University Press, Cambridge 2001;
Barth, T.: Error estimation and adaptive discretization methods in computational fluid dynamics. Berlin, Springer 2003;
Zienkiewicz, O., Taylor, R., et al.: Finite element method for fluid dynamics. Elsevier, Oxford;
Schneider, W.: Mathematische Methoden der Strömungsmechanik. Vieweg, Braunschweig 1978

[3.24] Komen, G. Cavaleri, L., et al.: Dynamics and Modeling of Ocean Waves. Cambridge University Press, Cambridge 2006, see also http://www.cambridge.org/0521577810;
Leal, J., Ferreison, R., et al.: Dam-break Wave-front Celerity. *J. Hydraulic Engrg.* 123, Nr. 1, January 2006, p. 69–76;

Mei, C.: The Applied Dynamics of Ocean Surface Waves. World Scientific, Singapore 1989, see also
`http://www.worldscibooks.com/engineering/0752.htm`;
Kowalik, Z.: Numerical Modeling of Ocean Dynamics. World Scientific, Singapore, 1993, see also
`http://www.worldscibooks.com/engineering/1970.htm`;
Gärtner, S.: Zur Berechnung von Flachwasserwellen und instationären Transportprozessen mit der Methode der finiten Elemente. VDI-Verlag, Düsseldorf 1977;
Holz, K.: Numerische Simulation von Flachwasserwellen mit der Methode der Finiten Elemente. Ein Beitrag zur Berechnung langperiodischer Flachwasserwellen nach dem Hamilton'schen Variationsprinzip. VDI-Verlag, Düsseldorf 1976;
Boccotti, P.: Wave mechanics for ocean engineering. Elsevier, Amsterdam 2000;
Koch, W.: Numerisches Charakteristikenverfahren unter Berücksichtigung von dissipativen Gliedern. Forschungsbericht 72-23, Deutsche Forschungs- und Versuchsanstalt für Luft- und Raumfahrt, Institut für Theoretische Gasdynamik, Aachen 1972;
Longmire, R.: Solution of Problems Involving the Hydromagnetic Flow of Compressible Ionized Fluids (characteristics method). *Phys. Rev.* 99, Nr. 6, September 1955, p. 1678–1681

[3.25] Gröbner, W., Knapp, H.: Contributions to the Method of LIE Series. Bibliographisches Institut, Mannheim 1967;
Wanner, G.: Integration gewöhnlicher Differentialgleichungen mit LIE-Reihen (mit Programmen). Bibliographisches Institut, Mannheim 1969;
Gröbner, W.: Über die Lösung von nichtlinearen Differentialgleichungen mit Randbedingungen. *Z. moderne Rechentechnik Automation* 9 Nr. 4, (1962), p. 147–151;
Gröbner, W.: Die LIE-Reihen und ihre Anwendungen. VEB Deutscher Verlag der Wissenschaften, Berlin, 1960 (Inversion of Functions), see also p. 64, 185 etc in [1.1];
Saely, R.: Development of New Methods for the Solution of Differential Equations by the Method of LIE Series, W. Gröbner, Contract JA-37-68-C-1199, European Research Office, July 1969;
Cap, F., Gröbner, W. et al.: Solution of Ordinary Differential Equations by Means of LIE Series. NASA Report CR-552, Washington, DC 1966

[4.1] Various public informations by the Hurricanes Research Division, available at
`http://www.aoml.noaa.gov/hrd/tcfaq;`
`http://www.nhc.noaa.gov/HAW2/english/storm`

[4.2] McGraw-Hill Encyclopedia of Science and Technology, New York 1960

[4.3] Pielke, R., Pielke, R. sen.: Storms, 2 vols. Routledge, London 2002; Foley, G., Willoughby, J. et al.: A Global View of Tropical Cyclones. World Meteorological Organization Report TCP-3P, 1995; Pichler, H.: Dynamik der Atmosphäre, 3rd ed. Spektrum-Akademischer Verlag, sections 8.11 and 9.2

[4.4] Pearce, R.: Why must hurricanes have eyes? *Weather* Vol 60, Nr. 1, January 2005; Pearce, R.: A study of hurricane dynamics using a two-fluid axisymmetric model. *Meteorol. Atmos. Phys.* 67, p. 71-81 (1998)

[4.5] Beaufort Scale in Encyclopedia Britannica, Vol. 3, p. 272, London 1960

[4.6] Bister, M., Emanuel, K.: Dissipative heating and hurricane intensity. *Meteorol. Atmos. Phys.* 50, 233–240 (1998);
`http://www.aoml.nooa.gov/hrd/tcfaq/G3.htm.`;
(Landsea C.: What may happen with tropical cyclone activity due to global warning?)

[4.7] Emanuel, K.: The dependence of hurricane intensity on climate. *Nature* 326, 483-485 (1987); Knutson, T., Tuleya, R.: Impact of $CO_2$-induced warming on simulated hurricane intensity and precipitation: Sensitivity to the choice of climate model and convective parametrization. *J.Clim.* 17, 3477–3495 (2004); Henderson-Sellers, A., et al.: Tropical cyclones and global climate charge: A post-IPCC assessment. *Bull. Ann. Meteorol. Soc.* 79, 19–38 (1998); Emanuel, K.: Increasing destructiveness of tropical cyclones over the past 30 years. *Nature* 436, Nr. 4, August 2005, p. 686–688

[4.8] `http://www.aoml.noaa.gov/hrd/tcfac/D7.htm`
(Landsea C.: How much energy does a hurricane release?); Austrian Newspaper "Der Standard", April 8, 2006, p. 35

[4.9] Peixoto, J., Oort, A.: Physics of Climate. American Institute of Physics, New York 1992

[4.10] Arfken, G.: mathematical methods for physicists. Academic Press, New York 1970; see also [1.1], [1.2]

[4.11] Ettling, D.: Theoretische Meteorologie. Vieweg, Wiesbaden 1996

[4.12] Exner, F.: Dynamische Meteorologie. Springer, Wien 1925

[4.13] Aberson, S.: Track and intensity models for hurricanes. In: `http://www.aoml.noaa.gov/hrd/tcfaq/F2.htm`

[4.14] Raethjen, P.: Dynamik der Zyklonen. Akademische Verlagsgesellschaft, Leipzig 1953

[4.15] Guldberg, Mohn: *Öst. Met. Z.* 12, 48 (1877), and Mohn, H.: Grundzüge der Meteorologie. Berlin 1879 (p. 216)

[4.16] Ertel, H.: Neuer hydrodynamischer Wirbelsatz. *Meteorol. Zeitschr.* (1942), p. 277

[4.17] Breitschuh, U., Jurisch, R.: Die Finite-Element-Methode. Akademie Verlag, Berlin 1993
Schwarz, R.: Methode der finiten Elemente. Teubner, Stuttgart 1980;
Gruber, R.: Finite Elements in Physics. North Holland, Amsterdam 1986;
Brebbia, C., Telles, J., Wrobel, L.: Boundary Element Techniques. Springer, Berlin 1983;
Törnig, W.: Numerische Mathematik für Ingenieure und Physiker. Springer, Berlin, 2 Bände, 1979;
Collatz, L.: THE NUMERICAL TREATMENT OF DIFFERENTIAL EQUATIONS. Springer, Berlin 1966;
Meis, T., Marcowitz, U.: Numerische Behandlung partieller Differentialgleichungen. Springer, Berlin 1978;
Marsal, D.: Die numerische Lösung partieller Differentialgleichungen in Wissenschaft und Technik. Bibliographisches Institut, Mannheim 1976

[4.18] Wacker, U.: Herberth, F.: Continuity equations as expressions for local balances of masses in cloudy air. *Tellus* 55A, 247–254, (2003)

[4.19] DeMaria, M.: Summary fo the NHC/TPC tropical cyclone track. report 11/26/97,
http://www.nhc.noaa.gov/modelsummary/shtml
Hurricane Tracking Software:
http://www.hurricane.com/hurricane-software.php

# Index

SpringerGeosciences

Bernhard Hofmann-Wellenhof,
Helmut Moritz

# Physical Geodesy

**Second, corrected edition.**
2006. XVII, 403 pages. 111 figures.
Softcover **EUR 59,–**
(Recommended retail price)
Net-price subject to local VAT.
ISBN-10 3-211-33544-7
ISBN-13 978-3-211-33544-4

"Physical Geodesy" by Heiskanen and Moritz, published in 1967, has for a long time been considered as the standard introduction to its field. The enormous progress since then, however, required a complete reworking. While basic material could be retained other parts required a complete update. This concerns, above all, the adaptation to the fact that the geometry can now be precisely determined by methods such as GPS, and that new satellite methods, combined with terrestrial methods, also make a detailed determination of the earth's gravitational field a possibility and a necessity. Highlights include: emphasis on global integration of geometry and gravity, a simplified approach to Molodensky's theory without integral equations, and a general combination of all geodetic data by least-squares collocation. In the second edition minor mistakes have been corrected.

P.O. Box 89, Sachsenplatz 4–6, 1201 Vienna, Austria, Fax +43.1.330 24 26, books@springer.at, **springer.at**
Haberstraße 7, 69126 Heidelberg, Germany, Fax +49.6221.345-4229, SDC-bookorder@springer-sbm.com, springer.com
P.O. Box 2485, Secaucus, NJ 07096-2485, USA, Fax +1.201.348-4505, service@springer-ny.com, springer.com
Prices are subject to change without notice. All errors and omissions excepted.

SpringerWienNewYork